THE SCIENCE OF SMALL ARMS BALLISTICS

THE SCIENCE OF SMALL ARMS BALLISTICS

Alvah Buckmore, Jr.
Ballistician & Theoretician

AAP | APPLE ACADEMIC PRESS

Apple Academic Press Inc.
3333 Mistwell Crescent
Oakville, ON L6L 0A2
Canada

Apple Academic Press Inc.
9 Spinnaker Way
Waretown, NJ 08758
USA

International Standard Book Number-13: 978-1-77188-650-5 (Hardcover)
International Standard Book Number-13: 978-1-315-14720-8 (eBook)

Library and Archives Canada Cataloguing in Publication

Buckmore, Alvah, Jr., author
The science of small arms ballistics : both for the academia and the recreational shooting community / Alvah Buckmore, Jr., ballistician & theoretician.

Includes bibliographical references and index.
Issued in print and electronic formats.
ISBN 978-1-77188-650-5 (hardcover).--ISBN 978-1-315-14720-8 (PDF)

1. Ballistics. 2. Firearms. I. Title.

UF820.B83 2018 531'.55 C2018-903197-2 C2018-903198-0

CIP data on file with US Library of Congress

Apple Academic Press also publishes its books in a variety of electronic formats. Some content that appears in print may not be available in electronic format. For information about Apple Academic Press products, visit our website at **www.appleacademicpress.com** and the CRC Press website at **www.crcpress.com**

I dedicate this book to my wife,

Lolita Buckmore

CONTENTS

ABOUT THE AUTHOR

Alvah Buckmore, Jr.
US Army Veteran, Small Arms Ballistics Expert

Alvah Buckmore, Jr., was a United States Army veteran during the time of the Vietnam War. When he came home to his parents in late 1974, after getting hurt very badly, he urinated with blood thick as tomato juice and weighed only 108 pounds. The Veterans Administration immediately took steps to nurse him back to health as an outpatient patient, and within four years, he had fully recovered from the weight loss and several infections. As part of his method of recovery, he went on a program of hunting, fishing, hiking, mountain climbing, target practice, reloading ammunition, and worked on a systematic study of the science of small arms ballistics.

He is a self-taught genuine expert on small arms ballistics with national and international recognition as a pioneer on this subject. He taught himself advanced mathematics and developed the skills and tools to work out the mathematical relationships of physical phenomena, starting off with the Theory of Transfer of Energy, Theory of Penetration, Field-Effect, and the Effect of Field-Effect Over Trajectory and Time. Among several other new theories in interior, exterior and terminal ballistics, he finally had enough material to write out an entirely original book on the science of small arms ballistics, working over a period of more than 20 years. He has also written and published on electronic radio communications and human sexuality.

LIST OF ABBREVIATIONS

APFSDS	armor-piercing fin-stabilized discarding sabot
BBL	best bullet length
BMV	best muzzle velocity
BPP	ballistic penetrating power
BRC	ballistic reflection coefficient
BRP	ballistic reflection power
BTU	British thermal units
CP	correct powder
ES	extreme spread
FN	flat-nose
GSW	gunshot wound
KE	kinetic energy
MKS	meter-kilogram-second system
MV	muzzle velocity
RCC	remaining case capacity
RE	reflection of energy
RN	round-nose
SD	standard deviation
SP	spitzer point
SPBT	spitzer point boat-tail
SVAT	soil-vegetation-atmosphere-transfer
SWC	semi-wad cutter
TE	transfer of energy
TOF	time of flight
UBP	uncorrected barometric pressure
WC	wad cutter

LIST OF SYMBOLS

Bd	bullet diameter
Bp	bullet density
P	density
V	velocity
MV	muzzle velocity
TE	transfer of energy
σ	air density
Æ	asymptotic function
Temp.	temperature
Þ	work function
g	acceleration due to gravity
ŵ	bullet spin
Ţ	terminal velocity
ƺ	initial terminal velocity
@	public domain
RCC	remaining case capacity
©	copyright
Lo	low
Hi	high
Av	average
ES	extreme spread
Sd	standard deviation
SD	sectional density
MER	maximum effective range
MRL	maximum range of lethality
MED	maximum effective distance
TW	twist
Bl	bullet length
d	diameter
BBL	best bullet length

BMV	best muzzle velocity
KE	kinetic energy
E	energy
D	distance

PREFACE

This book has the explicit purpose of serving as a manual for both members of academia and members of the recreational shooting community throughout the world. Though it contains extensive mathematical formulas (58 of them) in the English unit of measurement on the most part, the purpose of the book is not to intimidate anyone, regardless of his educational background, but to demonstrate the relationships between the physical variables and their real-world engineering applications. The formulas also have the explicit purpose to allow serious hunters and recreational and competitive shooters to reload ammunition with much greater precision and with much, much less guesswork.

Still a cursory examination of both the title and the table of contents, by anyone even remotely familiar with its literature will readily single out this book as very different from anything else on the market, either now or in the past. With very few exceptions, every book on the science of ballistics has been either confined to a tiny proportion of this subject, or it has been unreadable mathematically, with intensive gobbledygook written by someone with an advanced education on the science of mathematics, but who never took the time to communicate in the written word.

Even unclassified scientific material by government scientists on the science of ballistics is exceedingly difficult to read. We can find this material using any of the several search engines readily available on the internet.

Nor is it meant to be an intellectual abstraction of the science of small arms ballistics. Every word has an explicit purpose meant to be readable for the average person with an interest and desire to read and understand small arm ballistics with the purpose of developing a more solid grasp of this important but underappreciated science.

This book will define the science of ballistics, break it down into its major and logical categories and then explain the basic patterns and relationships in a way that, hopefully, the reader can understand and use toward the solution of his problems, either to design ammunition for a particular gun or a gun for any particular ammunition.

At each successive step, from interior to exterior to terminal ballistics, we will describe the major relationships that make up the science of small arms ballistics, including areas well outside of the science of ballistics. We will describe these relationships mathematically using standard systems of mathematical communications recognizable to the scientific community. In some instances, we will introduce new symbols and, in many instances, entirely new concepts, relationships, equations and even a demonstration of the potential for the development of new sciences, technologies and industries evolving from concepts and relationships in the science of small arms ballistics.

They will include the inception of the scientific study of time, evolving from the study of the field-effect theory in exterior ballistics, a recognition we may some day learn to manipulate space and time with the corresponding capability to travel through time from present to present and present to the future or past.

Though not an issue brought up in this book, it has become increasingly obvious to the perceptive person over the last century that we occasionally find visitors from other worlds, visiting us out of scientific curiosity, using a huge variety of advanced and very different if not entirely unique scientific technologies that we can easily perceive but is over our heads, including visitors from our own kind, from our own distant future, using time-travel as their mode of transportation. It has also grown increasingly obvious, as we study the cause and effect relationships in the science of exterior ballistics that some of these relationships can lead to the development of time-travel for us.

When we examine this subject in greater detail, it becomes just as obvious that this electrostatically charged plasma in outer space everywhere throughout the universe may have a memory attributed to it; meaning, it may be possible to use this plasma to store data in the same way we store data on a computer hard drive and, with this electrostatic charge, to expedite the movement of data flow.

Nor can we exclude the development of ballistic signatures, a way to identify the caliber, velocity and mass of a projectile in real-time flight, or to calculate trajectory backwards in order to identify a sniper's position.

There is even room to develop a science of force fields, evolving from the science of small arms ballistics, a science to absorb, reflect, or project kinetic energy as either a defensive or offensive weapon system.

Each equation, which describes a mathematical relationship, such as transfer of energy, will have an engineering application for the shooter with a design problem. Some equations, such as the calculation of bullet length with a given muzzle velocity and rate of twist, represent manipulations of those equations, such as the rate of twist and bullet spin. Some other equations represent a set of mathematical instructions to resolve a technical problem, such as the computation of trajectory or depth of penetration of living tissue in real-time.

Ballistics is a much bigger subject with a genuine need for much more thought and technical development than presently recognized by academia, recreational shooters, law enforcement, or the armed forces. It has the inherent potential to lead to the development of dozens of entirely new sciences, technologies, and vast industries most of us cannot even perceive at the moment.

—Alvah Buckmore, Jr.

INTRODUCTION TO THE SCIENCE OF SMALL ARMS BALLISTICS

A PREPARATORY HISTORY OF EARLY PHYSICS *BEFORE* WE GET INTO THE SCIENCE OF BALLISTICS

EARTH

Our Earth rotates on its own axis, in a clockwise position from west to east and hence the reason our Sun rises from the east and travels west. However, if we were in Outer Space a million miles or so from the planet Earth and looking straight down at it from the North Pole, it would look as if the Earth were rotating counter-clockwise from east to west. The axis of the Earth intersects in its surface at both the North and South Poles, with the Earth rotating around it once in every 24 hours around the Sun and, in

Photograph of Earth, taken during the Apollo 17 Lunar Mission in 1972. – Wikipedia, the free encyclopedia.

respect to the stars, once every 23 hours, 56 minutes and 4 seconds, but gradually slowing down due to the tidal effects of the Moon.

In the last 100 years, our Atomic clocks clearly show our Earth's days are longer by nearly 1.7 milliseconds and will get longer and longer every year until the end of our own Universe.

BRIEF HISTORY OF ITS SCIENTIFIC STUDY

Early man's attempts to scientifically study this subject was slow and exceedingly difficult, full of inconsistencies, extensive plagiarisms and frauds, and intimidations and contradictions in the interpretation of research data. He had to cope with the enormous complexity of looking at something very big, while very small, and without the proper support, material resources or technical instrumentation, such as a telescope. He also had to deal with the adverse intimidating side-effects of religion and culture over an interpretation of the available research information when it contradicted conventional intuitive wisdom or a belief system at the time.

Both religion and culture have always had the power to modulate man's interpretation of anything with the corresponding power to coerce him to seek a "confirmation" of the pre-dominate bias at the time. Objectivity was not only very difficult, in the best of times, but downright impossible and even dangerous to life and limb, such as when fighting the coercive powers of a powerful pagan religion, the Roman Catholic Church in medieval Europe or Nazi Germany in the 20th century. It takes more than intellect and integrity to be objective; sometimes it takes enormous courage, such as with Galileo (1564–1642) in his study of the astronomical observations and support for heliocentrism by Copernicus (1473–1543).

We will also see that when we study the various ancient Greeks in the Pythagorean School [note: *Pythagorean means of or pertaining to the ancient Ionian mathematician, philosopher, and music theorist Pythagoras* (c. 570–c. 495 BCE) – Wikipedia, the free encyclopedia], who believed the Earth rotated on its axis and around the Sun instead of the Universe (the Heavens) rotating around the Earth, they held a clear and direct contradiction to the opinions and conclusions of other, sometimes more politically or ecclesiastically correct, people much later, particularly when in contradiction to the Holy Bible and their scriptures

What we will find, in this history of the study of Earth's rotation and its relevance to the science of ballistics is the remarkable conclusion that, although these studies originated more than 2,500 years ago, there is still controversy and much more to learn about our planet.

More remarkable is the similarity between today's controversies of scientific research and development to those in more ancient times, particularly in certain areas of human endeavor. In that respect, nothing has changed!

Perhaps the first person in human history to have hypothesized the Earth rotates around its axis and around the Sun was Philolaus (470–385 BCE); but, according to August Bockh (1785–1867), Philolaus was the successor of Pythagoras, perhaps not the originator of this theory. We may never know for sure.

However, Philolaus's system of analysis and proof was difficult to follow with his "counter-earth rotating daily in the region of a 'Central Fire.'" He did away with the attitude of a fixed direction in space and

Bust of Pythagoras of Samos in the Capitoline Museums, Rome.

(Source: https://commons.wikimedia.org/wiki/File:Kapitolinischer_Pythagoras.jpg Creative Commons Attribution-Share Alike 3.0)

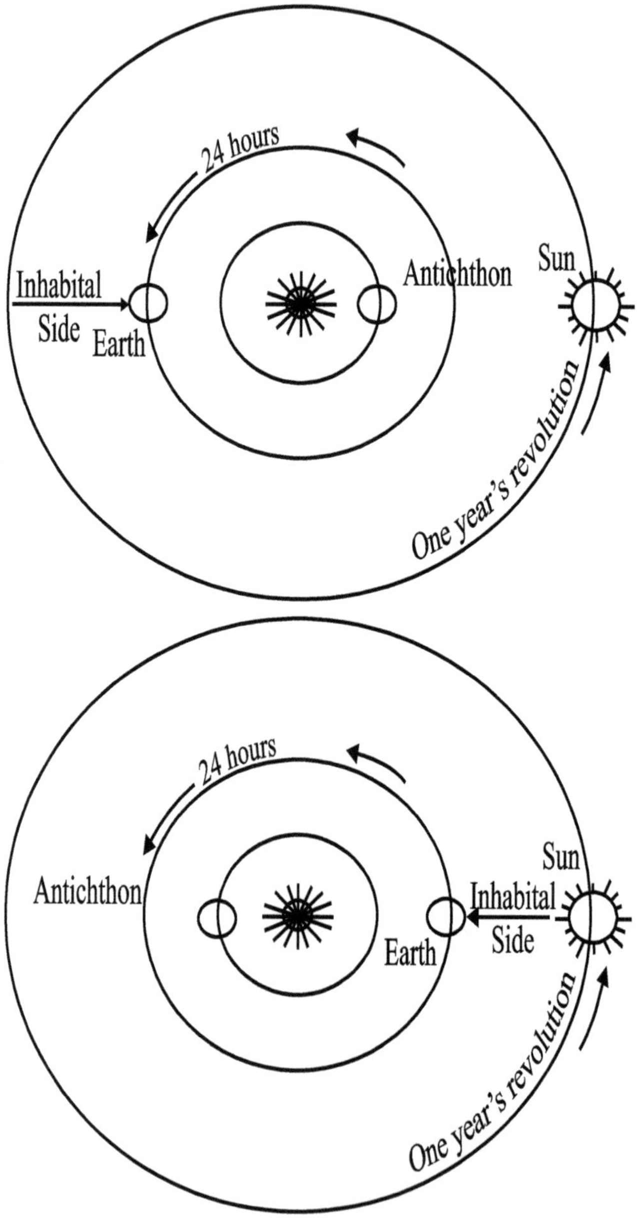

Philolaus, the successor of Pythagoras, and perhaps not responsible for the original theory. According to Stobaeus, his theory was not that the earth revolved around the sun (Heliocentrism), but that it revolved around a hypothetical astronomical object he called the "Central Fire," around which the sun also revolved around it.

instead developed one of the first non-geocentric views of the universe. He believed the Earth, Moon, Sun and other planets revolved around a hypothetical astronomical object he described as an unseen "Central Fire.

Perhaps, with his notion there is a "fire in the middle" of the Universe, he may have been thinking of the Sun as the *Central Fire* in the center of the Universe, though other accounts say otherwise (which may or may not be the results of confusion by Joannes Stobaeus, 5th-century CE from Stobi in Macedonia, who compiled a valuable series of extracts from Greek authors?)

Aristarchus of Samos (c. 310 – c. 230 BCE), an ancient Greek astronomer and mathematician, presented the first known model placing the Sun at the center of the known universe with the Earth revolving around it. Though influenced by Philolaus, he identified the "Central Fire" as the Sun, and put the other planets in their correct order of distance around the Sun.

When we look at the Earth from their vantage positions, looking at it without a telescope or any other instrument, or any prior real knowledge from earlier thinkers, we can easily recognize Philolaus's system as making a lot of sense, as well as with Aristarchus's perception.

Another controversy, supported by a variety of thinkers, such as Hicetas (400–335 BCE), Ponticus Heraclides (390–310 BCE) and Ecphantus (who may not have existed), accepted the notion or theory that the Earth rotates around its axis but not necessarily around the Sun.

Aristotle (384–322 BCE) was critical of Philolaus's ideas. He held the theory a sphere of stationary stars rotated around the Earth, a theory accepted by most people who followed him for centuries. Claudius Ptolemy (100–170 CE), as an example, believed, that if the Earth were rotating, as many people also thought at his time, then everything on it would violently fly off into space. That made sense without our knowledge of Gravity, a knowledge that came much later with the help by Galileo and Newton. Although Aristotle was clearly mistaken in his belief, it is equally clear he was conscientious in his efforts to teach himself, as well as his students, such as Alexander the Great, to learn to clearly and logically think, talk and to use information properly. In that respect, he was quite successful, definitely brilliant and immersed with the culture of war, conquest, slavery and subordination of people as an integral part of an economic system common in his days.

Aryabhata (476–550 CE) was an Indian astronomer who wrote in 499 CE that the Earth was round and rotated on its axis every day, and the movement of the stars was due to Earth's rotation. He used an analogy: "*Just as a man in a boat going in one direction sees the stationary things on the bank as moving in the opposite direction, in the same way to a man at Lanka the fixed stars appear to be going westward.*"

Some Muslim astronomers in the 10th century accepted the theory of Earth rotating around its axis and Abu Sa'id Ahmed ibn Mohammed ibn Abd al-Jalil al-Sijzi (c. 945–c. 1020 CE) held the notion "*that the motion we see is due to the Earth's movement and not to that of the sky...*" About that time, people who followed him also recognized "... *the earth is in constant circular motion, and what appears to be the motion of the heavens* (our Solar System) *is actually due to motion of the earth and not stars*." Nevertheless, for centuries after Philolaus, there were people who agreed and disagreed with him, including at least a dozen or more treatises written by them discussing the subject either accepting or rejecting Ptolemy's argument that, if the Earth were rotating, instead of the whole Universe rotating around the Earth (or the Heavens, as they understood it), everything on the planet would fly off in an enormous storm (or as gales, as they logically looked at it). All of this made sense, of course, when we lack knowledge of Gravity. Gravity kept everything from flying off, we eventually learned, thanks to Galileo and Newton.

Thomas Aquinas (1225–1274 CE), living in medieval Europe (between the 5th and the 15th centuries) at the time, accepted Aristotle's view on the subject "... that of a sphere of stationary stars rotated around the Earth ...," as well as scores of scholars, astronomers and philosophers, and not until did Nicolaus Copernicus (1473–1543) firmly establish the heliocentric world system, did the whole scientific community finally come around to accept the view that Earth rotates around the Sun instead of the whole Universe rotating around the Earth.

It took a long, long time, nearly 1,800 years in fact, for us to finally accept the Earth rotates around its axis and around the Sun and, now, with that basic knowledge, we can begin the study working toward the development of a real science of ballistics. Even then there was still controversy for another 100 years; or so.

We must realize it was very difficult for ancient man to study his world. Until fairly recently, man on the most part lived in acute poverty, health, nutrition and medical deprivation, and with an intense ignorance of the world around him. For thousands of years, he literally lived in a cave in the side of a mountain, a hole in the wall of a shelter, perhaps underground or some other primitive above-ground structure, and always without running water, heat or a means to keep himself clean or with an enough food to stay comfortable and healthy. Life was short!.

Everything was a mystery for him! Even a simple analysis or discovery of some physical or psychological phenomena took time to develop; sometimes, it took centuries for him to generalize from his observations and experiences to record them for the next generation. Though he certainly had the brain-power, it was not enough without an accumulation of real, verifiable knowledge, precisely the kind of verifiable knowledge that took literally hundreds – and indeed sometimes – thousands of years to accumulate along with the constant struggle of survival he faced every day.

Then, there was a constant challenge of his developing a suitable technology to record his growing accumulation of knowledge for the people who followed him. None of this was easy! And it is obvious now he had to learn and re-learn the same things over and over for thousands of years before it finally entered into a record for the next generation of thinkers. He needed to assess and assimilate what earlier man learned into more useful, verifiable knowledge and, at long last, a real, useful, usable and meaningful database for engineering applications and a better way to live. Nor can we discount the thousands of years of dangerous ideologies, religions, cults, pseudosciences, political power games of tyranny, slavery as an integral part of an economy, genuine conspiracies to obstruct intellectual development and even motivations out of jealousy.

We can go as recently as the Spanish Inquisition, Nazi Germany, the Stalinist Soviet Union or, as I write out this Chapter, the Taliban and insurgents in the Middle East who have destroyed the lives of millions of people out of religious intolerance and a quest for absolute power without responsibility.

There is an Islamic movement called the "Worldwide Caliphate" with the ultimate intent to "conquer the West" with a single theocratic one-world government proposed by some Muslims, such as *Abu Bakr al-Baghdadi*, described by the American government as the "ringleader of the Islamic State of Iraq and the Levant" (ISIL) (a term used synonymously with Syria-Palestine). He thinks of himself as the political and religious successor to the prophet Muhammad (c. 570–632 CE).

We can just as well imagine the lost of opportunity for development and genius with the death of so many innocent people with this kind of system of government, a system of government with a ruthless readiness and ambition of using advance military science and technology, such as modern artillery, machine-guns, jet fighters and nuclear physics (in order) to develop a nuclear bomb, while imposing a primitive life style, mores and standard of living for the common person common in the 6th and 7th century of the Common Era, with a primitive one- man theocratic rule, in the region we now call the Middle East. How long do you think that will last?

MEN WHO PILOTED THE WAY TO MODERN SCIENTIFIC THOUGHT

While we can successfully argue there were more than three men in antiquity responsible for such indisputable development of early scientific history – Socrates, Plato and Aristotle – it would be very difficult to find men with greater influence. These three men were the pivotal builders of modern science and, to a great extent, modern civilization today.

SOCRATES

Socrates (470/469–399 BC) was a classical Greek (Athenian) philosopher (by definition: a person who loves wisdom) may not have been the first but certainly one of the most significant founding fathers of modern scientific thought for the Western Civilization. Known largely through the accounts of classical writers more than 2,000 years ago, but more than ever through the writings of his students, Plato (428/427 or 424/423–348/347 BCE) and Xenophon (c. 430–354 BCE), Greek historian, soldier, mercenary and student of Socrates, and from the plays of the contemporary Aristophanes (c. 446–c. 386 BCE), a comic playwright of ancient Athens.

Socrates. (Source: Eric Gaba (User:Sting), July 2005. https://en.wikipedia.org/wiki/Socrates#/media/File:Socrates_Louvre.jpg. https://creativecommons.org/licenses/by-sa/2.5/deed.en)

Perhaps the most complete reports or accounts of Socrates to survive from antiquity came from Plato himself, though unclear as to the degree of which Socrates may have been involved in these written versions. Plato's dialogues made it clear, however, of Socrates' contributions to the field of Ethics, and bringing to our attention the *Socratic Method*, or *elenchus* ("*an argument that refutes a proposition by proving the opposite of its conclusions*" – Encarta Dictionary).

It remains a familiar tool in a wide variety of discussions (such as gun control), and is a type of pedagogy (*or concern in the study and practice of the best way to teach a person*). His method starts off with a series of questions – not only to extract individual answers – but to encourage his students to develop a fundamental insight into the subject through the process of participating in his dialog. Plato made it clear Socrates was also responsible for making important and permanent contributions to the field of Epistemology (*the branch of philosophy concerned with the nature and scope of knowledge and sometimes presented as a "theory of knowledge"*—Wikipedia, the free encyclopedia), and it remains a strong foundation for much of western philosophy and our scientific thought and endeavors today.

PLATO

Plato (428/427 or 424/423–348/347 BCE) was a philosopher and mathematician during the Classical Greece period (a 200-year period in Greek culture from the 4th and 5th centuries BCE) almost 2,400 years ago. He created the first institution of higher learning for us in the Western world and is widely recognized as the most important figure in the development of philosophy for us in the Western Civilization. Unlike nearly all of his philosophical contemporaries, Plato's entire work is believed to have survived intact for over 2,365 years, and he is possibly one of the greatest writers in the entire Western canon. The term – "Western Canon" – stands for a body of books, music and art accepted by Western scholars as responsible for shaping our Western culture. He is the pivotal figure responsible for our culture.

Alongside his teacher, Socrates, and his famous student, Aristotle, Plato was truly responsible for the very fundamental development of Western philosophy and science and, eventually, to the development of the science of ballistics. He was not only a pivotal figure for Western science, philosophy, and mathematics; he has also been named as one of the founders of Western religion and spirituality and, above all, Christianity.

Plato: copy of portrait bust by Silanion

ARISTOTLE

Aristotle, (which means "the best purpose"), was born in 384 BCE in Stagira, Chalcidice, about 55 km (34 miles) east of Thessaloniki (Salonika or Salonica, Greece), the second largest city in Greece and named after princess Thessalonike of Macedon, half sister of Alexander the Great.

Aristotle's father was Nicomachus (c. 375 BCE), the personal physician to King Amyntas of Macedon (died 370 BCE). We have very little real information about Aristotle's childhood. We believe, however, he probably spent some valuable time of his early life in the Macedonian palace, and developed his first important associations with the Macedonian monarchy at about the age of eighteen. He then moved to Athens to continue his education at Plato's Academy and, from there, remained for nearly twenty years before leaving Athens in 348/47 BCE.

Aristotle then accompanied Xenocrates (c. 396/5–c. 314/3 BCE) to the court of his friend, Hermias of Atarneus (originally a slave to Eubulus, a Bithynian banker who ruled Atarneus), in Asia Minor (who was also Aristotle's father-in-law). From there, he traveled with Theophrastus

Bust of Aristotle. Marble, Roman copy after a Greek bronze original by Lysippos from 330 BC.

(c. 371–c. 287 BCE), a Greek citizen of Eresos in Lesbos (the successor to Aristotle in the Peripatetic school; a school of philosophy in Ancient Greece) to the island of Lesbos, a Greek island located in the northeastern Aegean Sea, and, together, researched the botany and zoology of this island. See Sappho who was born between 630 and 612 BCE and died around 570 BCE.

She was a Greek lyric poet who lived on the island of Lesbos. Some historians have described her as one of the first known Lesbians in recorded history and, hence, many Lesbians today use the island of Lesbos as their Mecca, but not to the delight of the natives who live there. Some other historians have disputed the notion she was homosexual. They argue it was a horrendous misinterpretation of her poetry and wishful thinking on the part of some Lesbians looking for emotional support and historical justification for their homosexuality.

Aristotle married Pythias (died around 326 BCE), either Hermias's adoptive daughter or niece. She bore him a daughter, of whom they also named Pythias. Soon after Hermias' death, Aristotle was invited by Philip II of Macedon (382–336 BCE) to become the tutor to his son Alexander in 343 BCE.

Aristotle was appointed as the head of the royal academy of Macedon. During that time, he gave lessons not only to Alexander, but also to two other future kings, Ptolemy (c. 367–c. 283 BCE), a Macedonian general under Alexander the Great and ruler of Egypt (323–283 BCE) and founder of the Ptolemaic Kingdom and dynasty) and Cassandra (350–297 BCE), King of the Kingdom of Macedon from 305 to 297 BCE, and son of Antipater (c. 397–319 BCE) a Macedonian general, founder of the Antipatrid dynasty and, in his youth, educated by Aristotle at the Lyceum (a type of secondary school) in Macedonia.

FROM THE 10TH CENTURY ON . . .

For nearly 1,200 years, long after the death of Aristotle and his contemporaries, very little though and research was done by anyone concerning science and philosophy, and particularly the physics of our planet – Earth. In fact, as far as we can determine, outside of religion and the "proper" interpretation of the Christian Holy Bible, preciously few people did much thinking about anything else. Their whole lives evolved around their Bible, making it impossible for most of them to even consider anything other their immediate needs for survival, food for the winter, the harvest and family, along with an enormous emphasis on death and life after death (with very little *scientific* discussion, thought, debate or deliberation, if at all).

Long after the collapse of the Roman Empire, most people were simply too busy trying to find enough food to eat, fighting to stay alive and fighting each other. Not until the 10th century in the Common Era, did anyone do some serious thinking beyond food, sex and raw survival.

There were several exceptions, of course. During the 10th century, there were some Muslim astronomers who accepted that our Earth rotates around its axis. Al-Biruni (973–1048 CE) reported there was Abu Sa'id Ahmed ibn Mohammed ibn Abd al-Jalil al-Sijzi (al-Sinjari) (c. 945–c.1020 CE), an Iranian Muslim astronomer, mathematician, and astrologer, who argued "*... that the motion we see is due to the Earth's movement and not to that of the sky ...*"

He purportedly made an "astrolabe" (an instrument used to make astronomical measurements of the altitudes of celestial bodies, and in navigation for calculating latitude, before the development of the sextant), in a description detailed by al-Biruni, leading him to the conclusion the motion we can see is directly due to the Earth's rotation around its axis. Hence, the sky is not rotating around the Earth. It is the other way around.

As we can easily see now, it took a very long time for us to recover from the collapse of the Roman Empire. After al-Sinjari (see above), several important papers were written by some serious thinkers either to rebut or to express reservations above Ptolemy's argument that the Earth was rotating around its axis; if true, they argued, everything would have to violently blow off into Space.

Finally, in 1543 CE, Nicolaus Copernicus, a Renaissance mathematician and astronomer, just before his death, published his "Heliocentric World System" in his book entitled *On the Revolutions of the Celestial Spheres.*

NICOLAUS COPERNICUS

At the time of this publication, in 1543 CE, it was "...considered a major event in the history of science, triggering the Copernican Revolution and making an important contribution to the Scientific Revolution." – Wikipedia, the free encyclopedia

Nicolaus Copernicus.

... AND NOW WE ENTER THE AGE OF MODERN THOUGHT AND DEVELOPMENT...

Nevertheless, in spite of this "... major event in the history of science ...," we still have hundreds of years of academic struggle for freedom of thought and development before we can proceed beyond a superficial grasp of reality. Long before and after Nicolaus Copernicus, it was virtually impossible for most people to contradict the Roman Catholic Church; everything written, whether scientific or not, had to be in harmony with the "Sacred Scriptures." If not, then the writer was subject to severe criticism and punishment, including excommunications and even torture and death. Torture was the common tool to "persuade" him to admit to his errors. In other words, torture was used to obtain confirmation of what the Church wanted to hear, not to learn the truth.

We have that ugly situation today. One example would be the gun control controversy in the American news media. Throughout this industry, there is a strong "liberal bias" on the subject supporting gun control though, in spite of their contentions to the contrary, most news reporters and journalists know little or nothing about it. Hence, when an editor, news reporter, journalist or columnist receives an assignment to write on this subject, perhaps in response to a mass shooting incident in a school killing students and teachers alike, we will immediately find a strong pattern of the writer researching the subject in order to *confirm* his preconceived notions and arguments on the subject. We call this phenomenon confirmation bias.

Confirmation Bias

This pattern of *confirmation* is always the same as it was hundreds of years ago – no different from the confirmation patterns of the 10th to the 21st centuries. Only the period of time and the players differ.

In respect to the gun control controversy, the writer, thinking of himself as intelligent – and he is – and conscientious – in which he is not, will start off with a simple research program of accessing one of the search engines common on internet today. Typically, he will type in certain common key words in the address box of the search engine and, invariably, an unconscious bias will have led him to exactly what he wants to see, a *confirmation* of what he thinks he knows to be true, whether from various news organizations, anti-gun organizations and their articles and policy papers or the anti-gun statistics we find in their literature. Subsequently, his article will *replicate confirmation* of his bias and the bias of his work place, where he may have picked it up in the first place.

A research of Nicolaus Copernicus's professional history will reveal a similar pattern but a pattern of other people criticizing him for his failure to *confirm* their bias on the subject; for in this instance their bias is the attitude the Bible and its scriptures are always correct and anything that contradicts them are wrong and the work of the devil.

If we were to research Nicolaus Copernicus's personal history, as well as the personal histories of his contemporaries, we will find him – and his contemporaries – to have been exceptionally well-educated – even by today's standards. He spoke several languages, as did most of them; was a Renaissance mathematician and astronomer; a *polyglot* (a master of multiple languages) and *polymath* (a person who mastered several different subjects or bodies of knowledge); and who had obtained a doctorate in canon law while practicing as a physician, classics scholar, translator, governor, diplomat and economist. Even by today's standards, that is an enormous accomplishment. In the last 200,000 years, during the period believed our species, the Humo Sapiens, have existed, there is an estimated 60,000,000,000 men, women and children who have lived and died and, during that time, preciously few of them have ever reached the pivotal importance and accomplishments of Nicolaus Copernicus.

We will also find his enormous personal success was the direct results of political connections with important and politically powerful people along with the good fortune of having been born in a wealthy family. His family had the wealth and political connections to give him the opportunities to grow and to live with comfort few men and women will ever experience in a life-time, including political and ecclesiastical connections to the Roman Catholic Church. Those connections gave him power to influence people in power and the opportunity to research anything of interest to him, such as our Universe.

A study of this man's life, however, will lead us to the conclusion it was difficult for him, particularly during certain periods of his life, when he began to perceive contradictions between biblical scriptures and his observations of the physical universe physically observable to him (within the science and technology available at the time). For he had everything any man would want – wealth, power, education, family, etc. – everything except for one thing, the freedom to think.

A man in his position, with his close association with the Church and with all of his colleagues as members of the same Church, with the same kind of education centered on the Bible's *Scared Scriptures*, and the same set of prejudices and biases of the Church and the Polish *Intelligentsia* of the 16th century, it would not be just difficult but exceedingly dangerous for him to "think outside of that box."

As a Polish intellectual, scientist, scholar, etc., and his education controlled by the Church, he was placed inside of a box without the freedom to think outside of it. Any thought, theory or pronouncement in contradiction to this system of constraints could easily lead to his arrest, confinement, torture, death and even the entire destruction of his family. The pressure on him must have been absolutely enormous; hence, the reason his book, *Dē revolutionibus orbium coelestium* (On the Revolutions of the Celestial Spheres), was not published until the year he had died in 1543. Several other important people have had similar life-situations, such as Galileo.

We can find parallels of Copernicus's life predicament to accomplished people today, such as in the formal Soviet Union. Many years before its collapse, a powerful mayor of a large city in the Soviet Union, during a party with other powerful politicians and while participating in an informal conversation on economic and political reform in his country, woke up the following morning to find himself in a psychiatric institution with an enormous headache. He had been drugged in his drink the previous evening by someone who disapproved of his opinions. Now he must undergo "psychiatric treatment," his nurse told him, for having too many opinions contrary to the standard and accepted ideology at the time. He spoke "outside of the box."

Initially, there was very little controversy when Copernicus's book came out in print and without a fierce sermon, as was common when someone of this caliber became known to the *clerical intelligentsia* of those days. We are not sure of the reason it took nearly 60 years before the Catholic Church finally took an official position against it with their first step by the Magister of the Holy Palace (the Catholic Church's chief censor), Dominican Bartolomeo Spina (1475–1546), who wanted to "stamp out the Copernican doctrine."

However, with Spina's death, a well known theologian-astronomer, the Dominican Giovanni Maria Tolosani of the Convent of St. Mark in Florence (no personal information available), accepted the responsibilities, as his friend, to write out a denunciation of the *Dē revolutionibus orbium coelestium* before he saw a copy of it in 1544. His denunciation was written a year later, in 1545, found in an appendix of his unpublished work, *On the Truth of Sacred Scripture.*

Tolosani tried to disprove Nicolaus Copernicus's book using a philosophical argument and, all the while, calling upon Christian scriptures and traditions as his evidence. His objective was to demonstrate Copernicus's hypothesis of the Heliocentric World System as completely absurd. It lacked proof and clearly violated biblical scriptures, he argued.

"Nicolaus Copernicus neither read nor understood the arguments of Aristotle the philosopher and Ptolemy the astronomer," and added, he "… is very deficient in the sciences of physics and logic… *For it is stupid to contradict an opinion accepted by everyone over a very long time for the strongest reasons* [Author's emphasis], unless the impugner uses more powerful and insoluble demonstrations and completely dissolves the opposed reasons…"

John Calvin (1509–1564 CE), in his *Commentary on Genesis,* said, "We indeed are not ignorant that the circuit of the heavens is finite, and that the earth, like a little globe, is placed in the centre … For, just as in the wheels of a chariot there is an axle that runs through the middle of them, and the wheels turn around the axle by reason of the holes that are in the middle of them, even so is it in the skies…" *After all, "How could the earth hang suspended in the air were it not upheld by God's hand* [Again, the Author's emphasis]*?*"

THE PEOPLE WHO CAME BEFORE NICOLAUS COPERNICUS

At reported earlier, Philolaus described a "Central Fire" dominated the center of the universe, and a "counter-Earth" or "*Antichthon*" (a hypothetical body in the Solar system also purposed by Philolaus), with the Earth, Moon, the Sun and all of the planets, as well as all the stars, revolving around the "Central Fire." We do not understand the intellectual origin of his theories, however.

Heraclides Ponticus (387–312 BCE) suggested the Earth may rotate on its axis.

Aristarchus of Samos (310 BCE–c. 230 BCE) may have been the first to advance a theory that the earth orbited around the sun. The mathematics of Aristarchus' heliocentric system was worked out around 150 BCE by the Hellenistic astronomer Seleucus of Seleucia (c. 190 BCE - c. 150 BCE), an astronomer and philosopher. There is a reference in Archimedes' book of Syracuse (c. 287 BCE – c. 212 BCE), The Sand Reckoner (*Archimedis Syracusani Arenarius & Dimensio Circuli*), describing a work by Aristarchus of his advancing the heliocentric model. The original work by Aristarchus has been lost to history, unfortunately. Archimedes wrote:

"You (King Gelon) are aware the 'universe' is the name given by most astronomers to the sphere the center of which is the center of the Earth, while its radius is equal to the straight line between the center of the Sun and the center of the Earth. This is the common account as you have heard from astronomers. But Aristarchus has brought out a book consisting of certain hypotheses, wherein it appears, as a consequence of the assumptions made, that the universe is many times greater than the 'universe' just mentioned. His hypotheses are that the fixed stars and the Sun remain unmoved, that the Earth revolves about the Sun on the circumference of a circle, the Sun lying in the middle of the Floor [sic], *and that the sphere of the fixed stars, situated about the same center as the Sun, is so great that the circle in which he supposes the Earth to revolve bears such a proportion to the distance of the fixed stars as the center of the sphere bears to its surface."*–The Sand Reckoner (a work to determine the maximum number of grains of sand that could fit into the universe; written by Archimedes).

Copernicus cited Aristarchus of Samos in an early manuscript of *De revolutionibus orbium coelestium* (it still survives today), but removed the reference from his final published manuscript.

Since some of the technical details in Copernicus's Heliocentric World System closely bear a resemblance to those developed earlier by the Islamic astronomers, Naṣīr al-Dīn al-Ṭūsī (1201-1274 CE) and Ibn al-Shāṭir (1304–1375 CE), although we have no evidence Nicolaus Copernicus had read or had access to their works, it is easy for us to infer he did and may have used some of their material in his work, with or without a full consciousness of it. That is easy to do without awareness when we read thousands of pages of technical texts over a period of many years. We also have many instances of two or more people developing the same concepts or inventions at the same time while not even aware of each other's existence.

A case in point would be the development of Calculus in the 17th century Europe by Isaac Newton 1642–1726/7 CE) and Gottfried Wilhelm Leibniz (1646–1716 CE), with elements of this kind of mathematics appearing in ancient India, Greece, China, medieval Europe, and the Middle East over a period of at least 2,000 years.

In the early days of radio it was not uncommon for two people, unaware of each other, to invent the same thing with either slight or significant variations, and to introduce an application for a patent within a few days of each other. Of course, that would immediately create a lot of controversy and heavy litigation with suspicions and allegations of duplicity carrying throughout the entire radio communications industry to this very day.

Aryabhata (476–550 CE) (first of the foremost mathematician-astronomers from the classical age of Indian mathematics and astronomy), in his *Magnum opus Aryabhatiya* (499 CE), put forward a planetary model of the Earth spinning on its axis.

At the Maragha observatory (established in 1259 CE west of Maragheh, East Azerbaijan Province, Iran), Najm al-Dīn al-Qazwīnī al-Kātibī (died 1276/7 CE), in his "Hikmat al-'Ain," wrote an argument for a heliocentric model, but discarded it later. There is then a Qutb al-Din Shirazi, (1236–1311 CE), a 13th-century Persian polymath, poet and Arab Islamic astronomer, who made contributions to astronomy, mathematics, medicine, physics, music theory, Sufism (the study of the inner mystical dimension of Islam) and philosophy. He developed a geocentric system employing mathematical techniques, such as the Tusi couple (a small circle rotating inside of a larger circle twice the diameter of the smaller circle). Urdi lemma, born c. 1200 CE, probably in Urd, Syria, a major figure in 13th-century Islamic astronomy, was also almost identical in his efforts to those of Nicolaus Copernicus in the development of the heliocentric system in the 17th century.

PRELIMINARY SUMMARY

During Copernicus' lifetime, overshadowing all of the theories on this subject was Ptolemy's "Almagest," a treatise on the apparent motions of the stars and planetary paths, written and published around 150 CE; a hypothesis Earth was the stationary center of the universe with the stars

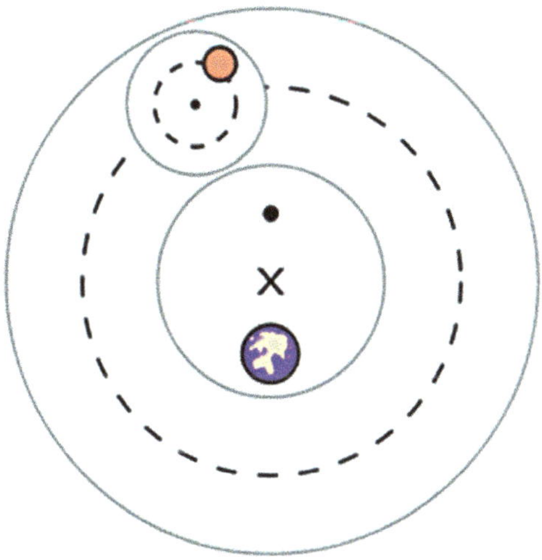

Ptolemaic model of the Universe with the Sun rotating around the Earth.

fixed in a large rotating outer sphere. His hypothesis was highly influential at the time and actually accepted as scientific fact for more than 1,200 years. Each of the planets, the Sun, and the Moon were also fixed in their own but smaller spheres. Ptolemy's system employed devices, including epicycles (a geometric model used to explain the variations in speed and direction of the apparent motion of the Moon, Sun, and planets), deferents (the condition of submitting to the superior influence of the objects in Outer Space) and equants (a mathematical concept developed by Claudius Ptolemy in the 2nd century CE to explain the observed motion of the planets), to account for the observations that the paths of these bodies differed from simple, circular orbits centered on the Earth. In other words, instead of perceiving the orbits of the Moon, Earth and planets as elliptical paths orbiting around the Sun, he perceived it as the Moon, Sun and planets as circular paths orbiting around the Earth, but aware of the contradictions in his calculations and unable to explain them in any other way, used epicycles, deferents and equants as a type of psychological rationalization to "explain away" these contradictions. Those systems of rationalizations in the support of his hypothesis continued for more than 1,000 years before

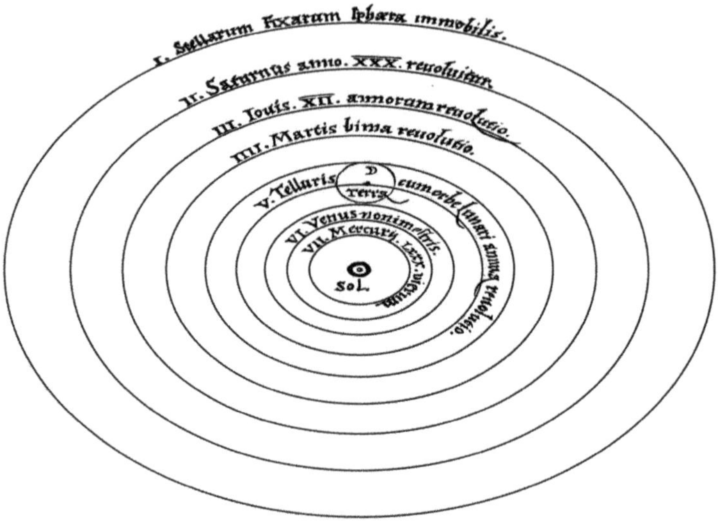

The above illustration is Copernicus's conception of the Universe. Notice, although the Sun is in the center of the Universe, all of the planets rotate around the Sun, which is correct, but he depicts them. As rotating around the Sun in complete circles instead of elliptical paths.

a trustworthy opposing argument and hypothesis by Nicolaus Copernicus in 1543 CE.

While some people have disputed the dates, Nicolaus Copernicus's *De revolutionibus orbium coelestium* has been recognized as the beginning of the scientific revolution in the Western Civilization from the year of 1543 of the Common Era.

THE PEOPLE WHO CAME AFTER NICOLAUS COPERNICUS

There were several distinctive phases in this scientific revolution started by Copernicus; --initially, the first phase focused on the recovery of the knowledge of the ancients, long forgotten by most people at the time, and has been described as a *Scientific Renaissance.* With the publication of Galileo's Dialogue Concerning the Two Chief World Systems in 1632 CE, this Scientific Renaissance was said to have ended starting with a new age. The completion of the scientific revolution is attributed to the "grand synthesis" by Isaac Newton's 1687 *Principia*, that formulated the laws of motion and universal gravitation. By the end of the 18th century, the scientific revolution had given way to the *Age of Reflection*.

The concept of a scientific revolution taking place over an extended period emerged in the eighteenth century with the work of Jean Sylvain Bailly (1736–1793), who saw a two-stage process of sweeping away the old and establishing the new. He was a French astronomer, mathematician, freemason and political leader in the earlier phase of the French Revolution He served as the mayor of Paris from 1789 to 1791, but guillotined during the Reign of Terror (from September 5, 1793 to July 28, 1794) for refusing to testify against Marie Antoinette.

Then it gets a little complicated. Although Nicolaus Copernicus's heliocentric theory was generally accepted almost universally in Europe, there were only about 15 astronomers actively promoting his theory. Georg Joachim Rheticus (1514–1574) had the potential of becoming Copernicus's successor if it were not for his crimes. In April 1551, he was accused of raping the son of Hans Meusel, a merchant. According to *Jack Repcheck, in his Copernicus's Secret: How the Scientific Revolution Began*, Rheticus allegedly "plied him with a strong drink, until he was inebriated; and finally did with violence overcome him and practice upon him the shameful and cruel vice of sodomy [an offensive term for anal intercourse of children, adults or animals]". A year later, he was found guilty and banished from Leipzig for 101 years.

Erasmus Reinhold (1511–1553), unfortunately, died prematurely at the young age of 42. A colleague of Georg Joachim Rheticus, he catalogued a large number of stars but rejected the heliocentric cosmology (the study of the origin, evolution and future of the universe) by Copernicus on physical and theological grounds. With that rejection, he was committed to the Biblical scriptures and Ptolemy's *Almagest* (a mathematical and astronomical treatise on the apparent motions of the stars and planetary paths written in the 2nd century CE).

Apparently, the first of the really great successors to Nicolaus Copernicus was Tycho Brahe (1546–1601), but did not believe the Earth rotated around the Sun. He was well known during his lifetime as an astronomer, astrologer and alchemist and, recently, described as a highly competent mind in modern astronomy who held a passion for facts derived empirically; yet, he also proved unable to separate these facts from Biblical scriptures and Ptolemy's *Almagest* written nearly 1,400 years earlier.

TYCHO BRAHE

There were people, such as Thomas Digges (no information available) and Thomas Harriot (1560–1621) of England; Giordano Bruno (1548–1600) and Galileo Galilei (1564–1642) of Italy.

Diego Zuniga (1536–1597) was from Spain.

Simon Stevin (1548–1620) worked as a Flemish mathematician, physicist and military engineer of the Low Countries.

In Germany, there was an exceptional group of successors, such as Georg Joachim Rheticus (1514–1574), known for his trigonometric tables and as Nicolaus Copernicus's sole pupil; Michael Maestlin (1550–1631) a German astronomer and mathematician, known for being the mentor of

Tycho Brahe (1546 to 1601).

Johannes Kepler; Christoph Rothmann (born between 1550 and 1560 and died probably sometime after 1600 and who may have later recanted on his views), and Johannes Kepler (1571–1630), a German mathematician, astronomer, and astrologer, and certainly an crucial person in the 17th century scientific revolution.

We have some additional possibilities: The Englishmen William Gilbert (1544–1603); Achilles Gasser (1505–1577), a German physician and astrologer and supporter of both Copernicus and Rheticus; Georg Vogelin Otto (no information available) and Tiedemann Giese (1480–1550), a proponent of heliocentrism by Nicolaus Copernicus.

The intellectual atmosphere lingered for years dominated by Aristotle's work and the Ptolemaic astronomy. Few were able to accept the Nicolaus Copernicus heliocentric theory. It violated the Church's scriptures and more than 1,500 years of precedence; even today, few men would have the emotional and personality skills and tools to contradict such power.

Tycho Brahe's theory, in spite of his enormous empirical work and accomplishments, held the theory our Earth is stationary, with the Moon revolving around it; the Sun revolving around the Earth and all of the other planets revolving around the Sun. That, of course, directly contradicted Copernicus's heliocentric theory.

About 50 years later, Kepler's and Galileo's work provided a substantial and solid evidence supporting Copernicus's theory, probably originating when Galileo first formulated the principle of inertia to explain the reason everything would not fall off if the Earth were in motion, and when Isaac Newton developed a universal law of gravity and the laws of mechanics in his 1687 book, *Principia.* That led to the development of a unified terrestrial and celestial theory of mechanics. From there, there began a general acceptance of Nicolaus Copernicus's Heliocentric World System.

In the Copernican system, the Earth and other planets orbited the Sun; while in the Ptolemaic System, everything in the Universe orbited the Earth.

JOHANNES KEPLER

Johannes Kepler (1571–1630) was a German mathematician, astronomer and astrologer, and an important figure in the scientific revolution of the 17th century. Best known for his laws of planetary motion, he built them on his works in *Astronomia nova, Harmonices Mundi* (The Harmony of the World) and *Epitome of Copernican Astronomy*. They represent a crucial foundation for Isaac Newton's later theory of universal gravitation and, without Kepler's work, it will have been much more difficult for Newton to have developed such a theory. He later became an assistant to Tycho Brahe and the imperial mathematician to Emperor Rudolf II. He did work in the field of optics and invented an improvement of the refracting telescope we now call the Keplerian Telescope.

Not a true scientist by today's standard, motivated and driven by his religious conviction and his infallible belief in God; he incorporated religious arguments and logic into his scientific work. To him, God created the world in harmony to an intelligible plan we could access through the light

Johannes Kepler (1571–1630).

of reason; his work confirmed it, he reasoned. Even though an integral part of the Scientific Revolution of the 17th century, he was nevertheless not a true scientist; however, his work made a most significant contribution to the development of modern science, a contribution we cannot ignore.

GALILEO GALILEI

Galileo Galilei (1564–1642), an Italian astronomer, physicist, engineer, philosopher and mathematician, played a major role in the *Scientific Renaissance*. His achievements include improvements to the telescope and the subsequent astronomical observations and support for heliocentrism. Galileo has been called the "father of modern observational astronomy," the "father of modern physics," and the "father of modern science."

He contributed significantly to observational astronomy plus the confirmation of the phases of Venus, the discovery of the four largest moons of Jupiter and the observation and analysis of the sunspots. He even invented an improved version of a military compass as well as other instruments.

Galileo Galilei (1564–1642).

His support of Nicolaus Copernicus's heliocentrism led to a great deal of controversy throughout his life at the time most people accepted either Aristotle's Geocentrism or the Tychonic system.

This whole matter was investigated by the Roman Inquisition in 1615. They came out with the conclusion that it could be a possibility but not as a reputable fact.

He defended his position in a book entitled, *Dialogue Concerning the Two Chief World Systems*, and alienated Pope Urban VIII when it appeared to have attacked him.

The Inquisition tried him for being "vehemently suspect of heresy;" he was to recant and required to spend the rest of his life under "house arrest." It was then he wrote perhaps his finest scientific work, the *Two New Sciences*.

After his publication of *Dialogue Concerning the Two Chief World Systems*, the Roman Inquisition banned the publication of all of his works, including books he might write in the future. It did not seem to matter whether or not any following material in the future could or would violate Biblical scriptures. They apparently assumed the worse.

After the failure of his initial attempts to publish *Two New Sciences* in France, Germany and Poland, it was published in Leiden, South Holland, where the writ of the Inquisition produced fewer adverse effects. Fra Fulgenzio Micanzio (1570–1654), the official theologian of the Republic of Venice, had, at first, offered to help Galileo publish his book in Venice, but pointed out that publishing the *Two New Sciences* in Venice might cause Galileo unnecessary trouble; however, the book was, in spite of this potential trouble, published in Holland. Yet, Galileo did not seem to suffer any harm from the Inquisition for publishing this book given that, in January 1639, the book reached Rome's bookstores with about 50 copies quickly sold. On the face of it, everyone liked it.

His opponents sometimes used the biblical excerpts of Psalm 93:1, 96:10, and 1 Chronicles 16:30 to attack heliocentrism. That included texts stating that "the world is firmly established, it cannot be moved."

In Psalm 104:5 it says, "The Lord set the earth on its foundations; it can never be moved."

What's more, Ecclesiastes 1:5 says that "... And the sun rises and sets and returns to its place." Galileo argued heliocentrism did not contradict biblical texts.

By 1615, Galileo's submitted his material on heliocentrism to the Roman Inquisition; but his greatest offense, from the view of the Roman Catholic Church, was his efforts to reinterpret the Bible, viewed as a violation of the Council of Trent and dangerously resembling Protestantism. Only the Church can interpret the Bible.

An Inquisitorial commission, in February 1616, affirmed heliocentrism to be "foolish and absurd in philosophy, and formally heretical since it explicitly contradicts in many places the sense [sic] of Holy Scripture."

Pope Paul V instructed Cardinal Bellarmine to present this decision to Galileo and to order him to abandon his opinion heliocentrism was physically factual. Then, on February 26, Bellarmine called Galileo to his residence and ordered him to abandon his opinion that the sun is stationary at the center of the world while the earth moves around it and, from this time forth, not to hold, teach, or to defend it in any way, either orally or in the written word.

The sentence of the Inquisition was delivered on June 22. It consisted of three parts:

Galileo was found "vehemently suspect of heresy," namely of having held the opinions that the Sun lies motionless at the centre of the universe, that the Earth is not at its centre and moves, and that one may hold and defend an opinion as probable after it has been declared contrary to Holy Scripture. He was required to "abjure, curse and detest" those opinions.

He was sentenced to formal imprisonment at the pleasure of the Inquisition. On the following day this was commuted to house arrest, which he remained under for the rest of his life.

His offending Dialogue was banned; and in an action not announced at the trial, publication of any of his works was forbidden, including any he might write in the future.

Allegedly, after recanting his theory that the Earth moved around the Sun, he muttered the rebellious phrase, "*And yet it moves*." He was referring to the Earth moving around the Sun.

ISAAC NEWTON

Sir Isaac Newton (1642–1726/7), an English physicist and mathematician, but in his time described as a "natural philosopher," was widely acknowledged as one of the most prominent scientists and as a central figure in the *scientific revolution*. His book *Philosophiæ Naturalis Principia Mathematica* ("Mathematical Principles of Natural Philosophy"), and first published in 1687; laid the fundamental principles for the development of classical mechanics. Newton made vital contributions to the science of optics; and it is now clear both he and Gottfried Leibniz (1646–1716) invented calculus independently of each other and almost at the same time; although at that time, many people did not think so. It created an enormous controversy among the contemporary mathematicians and gradually destroyed the relationship between the two men.

Isaac Newton (1642–1726/7).

Newton's law of universal gravitation

Newton states "that any two bodies in the universe attract each other with a force that is directly proportional to the product of their masses and inversely proportional to the square of the distance between them," a general physical law of gravity of paramount importance he derived from empirical observations. Of course, we cannot neglect the work of several other physicists, such as Robert Hooke (1635–1703), actually also an English natural philosopher, architect and polymath, who, in a communications with the Royal Society in 1666, slowly drew similar conclusions but without the mathematical hypothesis of the gravitational attraction "inversely proportional to the square of distance" between the "celestial bodies" in the Universe.

Newton's work eventually became an integral part of classical mechanics he formulated in *Philosophiæ Naturalis Principia Mathematica* ("the *Principia*"); he first published on July 5, 1687. It is now obvious, after a study of this subject and Newton's work; Newton did not develop his laws of universal gravitation in a vacuum. Several other people, including Robert Hooke, made significant contributions to influence his hypothesis, as would be common with all of us when the time is ripe for such intellectual activity. No one else works in a vacuum, either!

Gravity or gravitation has been now recognized as a "natural phenomenon" in the Universe in which everything with mass "gravitates" toward each other and that it is mass that makes up everything in the Universe – stars, planets, galaxies, light and even sub-atomic particles.

Gravity even creates spheres of hydrogen and ignites them under intense pressure to form stars and then groups them into galaxies.

It is gravity that makes our Universe what it is and gives it weight to everything else if it contains mass. Gravity on the Moon even causes the tides on our planet, but while not subject to absorption by or conversion to anything else. Nor can we, at least presently, remove or shield ourselves against gravity. It is always present if there is mass.

WHAT HAS ALL OF THIS LED TO?

At this point in time, when Newton had worked out a theory of gravity, within a very short period of time it led to the development of other theories of the physical world, most notably the study of light, energy, kinetic energy, mass, weight, velocity, acceleration due to gravity, drag, the recognition of the parabolic trajectory, the effect of the Earth's rotation over the deflection of moving projectiles, etc.

All of these studies, and many others, were absolutely necessary before it could lead to the modern development of the science of ballistics, its subsequent engineering applications and scientific outshoots, including the critical need for the development of calculus by Newton and Leibniz. Calculus made this Science workable by providing a method to calculate trajectory, depth of penetration or the effect of field-effect over time, trajectory or gravity, etc – all in real-time we can now easily do with high-speed digital computers.

Until recently, without such computers, even with advance calculus, there was no such capability. Slide rulers and manual computations were simply not adequate, and far too slow, to handle these problems due to the huge number of iterations.

Even with a simple computation of the transfer of energy, without taking into consideration of the other two components of drag, computing in increments of cubic inches for small-arm ballistics, there are no less than 3,600 iterations of computation for every interval of 100 yards. Until the development of the 80186 micro-processor, it would take at least 30 minutes to compute out to every 100 yards; impossible for real-time requirements; then that the 80186 micro-processor took almost 20 minutes.

Finally, with the 80486, the speed was dropped down to approximately nine seconds, still nowhere near real-time computations for anything realistic out to anywhere near 1,000 yards for small arms, and certainly nothing we can use for field artillery. Up to that time, all computations were approximate and very inaccurate if we were striving for precision, usually calculating in intervals of 100 yards. Computing in intervals of 100 yards, by the way, instead of cubic inches could easily cause inaccurate results in excess of 40 percent when computing out to 500 yards, a totally unsatisfactory solution to this problem.

Then that would only apply to the calculation of the loss of kinetic energy and velocity through transfer of energy. It would not include the other two components of drag or the algorithms to include the calculation of a bullet's trajectory above or below baseline in small arm ballistics.

Analog computers took much longer, actually several hours in the days when each computer came with its own individual operating system and language and could easily cost a million dollars a piece with follow-up logistics and maintenance programs by the manufacture, making real-time computations even much more unlikely. The technology was simply not available.

Until the development of such computers became available, this Science was literally on "hold." Just before and during the Second World War, analog computers were in use for the calculation of a projectile's trajectory in field artillery, coastal artillery and ship-to-ship artillery, and to calculate for the compensation of the coriolis effect (the calculation of the projectile's deflection, due to Earth's rotation on its axis, to the right in the Northern Hemisphere or to the left in the Southern Hemisphere) (very important when dealing with distances substantially beyond 1,000 yards). Or, else, we would not hit an object the size of an aircraft carrier at 10 miles with a 2,000 pound projectile.

Then, about the same time, there was also an urgent need for computers to handle the critical computations in nuclear physics when working toward the development of the first Atomic bomb in the 1940's. Yet, they were still far too slow for real-time computations and, in many respects that is still true today even with the advance processors presently on the market. We are still waiting for other more advance developments in smaller packages for real-time small arm ballistic computations out in the field.

During the Second World War, the British intelligence community developed an electrical/mechanical computer they called the "Bomb" to decode German secret messages they intersected from their radio transmissions. Initially, it took several days to decode one message from German Headquarters to a command out in the field. Then, as the British decoding mathematicians improved their grasp of the protocol and procedures used by German cryptographers, they rapidly developed the algorithms to increase the speed of decoding messages to several minutes.

The Americans had similar problems with the development of the first nuclear submarine, the Nautilus, during the 1950s. When President Eisenhower had ordered the U.S. Navy to send the Nautilus on an expedition under the polar ice in the North Pole, they soon discovered the lack of sufficient "computing power" a serious handicap. In those days, computers were still mechanical or electrical/mechanical monstrosities far too big and heavy to put inside of a submarine. They were necessary for both navigation and the safe control of the nuclear reactor which were, at that time, operated manually for lack of "computing power. "

As we can see, it was not just the lack of "computing power" before we could get to this level of technical development, it took nearly 1,200 years for someone to perceive, understand and to work out the mathematical relationship of gravity. Without it, there can be no science of ballistics. Access to a powerful computer will not have been enough; however, it is absolutely essential in order to apply the mathematics as we shall see in the following chapters.

For a real science of ballistics to develop, then again, we needed to understand the Earth is round, rotates on its axis and rotates around the Sun. Then, we needed to perceive, recognize and to work out the mathematical relationship of gravity, acceleration due to gravity (for the calculation and prediction of a bullet's trajectory, as one example; or to calculate a bullet's weight, as another example); aerodynamic drag, drag induced by the surface area of the nose, drag induced by the surface area of the cylinder, transfer of energy and deflections of falling bodies.

We needed to understand this effect of gravity over everything on our planet for us to understand things will not fly off into space due to the rotation on its axis, as did Claudius Ptolemy thought in the 2nd century of the Common Era.

Still, we had people, such as Giovanni Battista Riccioli (1598–1671), who disputed Copernicus's model of the Earth rotating around its axis, nearly a century after his death, due to his inability to observe the eastward deflections of falling bodies recognized today as the coriolis effect (see the Chapter on the coriolis effect) in the Northern Hemisphere.

Algebra, another critically important development, can be traced to the ancient Babylonians (historically, a period between 1800 and 1600 BCE, the kind of information we derived from their clay tablets) all the way up to the present, now with at least 21 different developed fields.

Geometry, a science of mathematics originating from Ancient Greece, is a branch concerned with questions of shape, size, relative position of figures and the properties of space, all of it important in the calculations of a projectile's physical parameters, either in small-arm ballistics or any projectile in free-flight. As an example, we would need to use geometry to aide in the calculation of the bullet's drag induced by the surface area of the nose and drag induced by the surface area of the cylinder.

When we study this history, leading to the development of the science of ballistics, we must take into account the enormous difficulties of every person in this chain of development, including people who did not make contributions for whatever reason, but came close to it.

From Aristotle's time to the present, over a period of more than 2,500 years, every person, successful or not, had his own problems, sometimes insurmountable. He had to start from nothing, originally with no support, material comfort or technical resources. All he had was his brain with an intellectual curiosity, desire and capability to understand his world around him. It was not easy; – in fact, for most people, it was impossible. Hence, very, very few people made such a notable gain toward the development of a real insight and knowledge of physical reality. During that same period of time, long before it and certainly to the present, most of this development was nonsense and sometimes dangerous. It led to many false paths and to the destruction of countless lives through intolerance and sheer ignorance, sometimes jealousy and all too frequently wars.

Throughout our entire human history, without a single exception, we have been in a constant state of war somewhere on this planet and, with preciously few exceptions; every one of them started by someone, or a group of people, motivated and driven by a quest for power, the kind of power without responsibility, transparency or accountability.

When we look back at world history, as far back as we can, from pre-history to our present history, it has been a history of people striving for advantages, privileges and power over everyone else. From before the time of ancient Babylonia, to the Persian Empire, to the Roman Empire, to the Ottoman Empire, to the recent particularly destructive Nazi German Empire, not excluding the primitive dynasties in ancient Egypt, there was very little permanency in anything they did, except for the permanency in the history of their existence.

Ancient China was constantly in war with each other and constantly driving their enemy tribes west of them further west into Eastern Europe and, finally, to Western Europe, to continually challenge the authority, privileges and power of the Roman Empire, an empire with an economy based on slavery (as was true with most empires) (with the British empire as an important exception), until this Empire finally collapsed from too many enemies, too many battles and the exhaustion of their resources and manpower.

Nor can we exclude the ancient primitive empires in both Latin America and South America. They were no different! Their economies were also based on slavery while stealing wealth from their neighbors, subsequently making real scientific and material development almost impossible – and sometimes completely impossible.

This pattern has always been the same. People in a position of power use their position of power to exploit the availability of resources outside their immediate sphere of influence in order to expand their sphere of influence. As in the case of ancient Rome, but equally true with every other empire, once the person or people in power had completed the consolidation of power, the immediate objective was always to expand into other territories (look at Nazi Germany and the Empire of Japan as recent examples). With superior military power, they would conquer the land outside of their own sphere of influence by killing the men and women who fought against them and then reducing the remaining population into slavery and poverty, and all the while stealing their wealth and material resources. This always had the effect of destroying careers, lives and opportunities; hence the reason it took so long to get this far and, if that were not enough, we ought to consider the colossal number of people who have died prematurely from disease.

INTRODUCTION IN THE SCIENCE OF SMALL ARMS BALLISTICS

*"**Ballistics,*** [is the] science of the propulsion, flight, and impact of *projectiles. It is divided into several disciplines. Internal and external ballistics, respectively, deal with the propulsion and the flight of projectiles. The transition between these two regimes is called intermediate ballistics. Terminal ballistics concerns the impact of projectiles; a separate category encompasses the wounding of personnel.*"

—Encyclopedia Britannica

*"(General Physics) (functioning as singular) the study of the flight dynamics of projectiles, either through the interaction of the forces of propulsion, the aerodynamics of the projectile, atmospheric resistance, and gravity (**exterior ballistics**), or through these forces along with the means of propulsion, and the design of the propelling weapon and projectile (**interior ballistics**)"*

—The Free Dictionary by Farlex

DEFINITION

Ballistics – the 61st branch of the science of Physics – starts from particle physics, which deals with the study of sub-atomic particles travelling at or very near the speed of light and stops at the physics of time, presently a non-existent science which, if it were to exist, would deal with the study of the relationships and the phenomena of light, gravity, electromagneticism, etc., responsible for the expansion and contraction of time. Evolving as other offshoots would be scores of new sciences, such as the science of Ballistic Signatures or the science of Force-Fields.

Ballistics, specifically, deals with the study of relatively large projectiles travelling at relatively slow velocities and always substantially below the speed of light and always in free-flight.

Its study starts at the precise moment of ignition of the propellant; to the study of the patterns and relationships of the projectile travelling through the bore; to the calculations and predictions of flight and finally to the study of the patterns and relationships between the projectile in flight, the terminal flight in the target and the target itself.

Traditionally, Ballistics has been divided into three major categories: Interior Ballistics, Exterior Ballistics and Terminal Ballistics.

- *Interior Ballistics* contains several topics, concepts, theories and relationships never before in print. Starting with the characteristics and instructions for reloading ammunition, it goes into Statistics, Kinetic Energy and the Theory of Twist with equations to calculate the best rate of twist, when designing a barrel; best muzzle velocity for a given bullet and rate of twist, when we must work with a given barrel and twist; best bullet for a given muzzle velocity and rate of twist, when we can design a bullet for a given muzzle velocity and rate of twist; and the Theory of Spin and Bullet Geometry.

- *Exterior Ballistics* will also contain several new topics, concepts, theories and relationships; including The Field-Effect Theory; The Effect of Field-Effect Over Time; The Effect of Gravity Over Time; The Effect of Field-Effect Over Trajectory; and an highly advance method to calculate trajectory in real-time with the algorithms to calculate the Maximum Range of Lethality and Maximum Effective Range.
- *Terminal Ballistics* will be no exception, either. It will contain the Theory of Transfer of Energy; Acceptance of Energy; Reflection of Energy; a Theory of Penetration and the algorithms to calculate depth of penetration into living tissue, and the range of lethality beyond penetration. Indeed, we sincerely hope and desire that this entire book should represent an enormous advancement for the science of ballistics and certainly for all members of every community.

SECTION ONE

CHAPTER 1

THE SCIENCE OF INTERIOR BALLISTICS

CONTENTS

1.1 DEFINITION

Interior ballistics is the scientific study of the physical phenomena that occurs inside and outside of the receiver and barrel of a gun.

Its study starts at the precise moment the firing pin strikes the primer and stops at the precise moment the bullet leaves the barrel.

It includes a study of the effects and characteristics of gunpowder at the precise moment of ignition and continues until the bullet leaves the barrel.

It includes the study of the bullet's spin and the barrel's rate of twist.

It also includes the study of the effects and characteristics of the outside physical environment over the environment inside of the receiver and barrel.

1.2 THE PROBLEMS AND METHODS OF DESIGNING AND RELOADING AMMUNITION

In conventional reloading circles, the normal procedure to reload ammunition is to get a simple reloading press, the appropriate dies for the caliber and cartridge, a pile of new or spent cartridge cases, some primers, and perhaps a pound or two of gunpowder. Then, with all the necessary equipment and material in hand, we would study the first available "reloading handbook" or "manual" for a "recommended load."

All gunpowder and most bullet manufacturers provide a manual to list their "recommended loads" for each cartridge. Some of these manuals are enormous books in actual physical size, and others represent nothing but a sheet or two of paper with some data. Each gunpowder manufacturer provides a comparison chart to allow the reloader to compare the burning rate characteristics and time-pressure curves of one particular powder to another particular powder, usually starting with the fastest burning powder on the left or top of the sheet of paper and moving to the slowest burning powder to the right or bottom of the paper. Then, starting with the smallest caliber, they would list a range of minimum and maximum recommended loads for each gunpowder suitable for each cartridge of each caliber and bullet weight.

Some gunpowder manufactures only recommend one load for each powder and bullet combination for each cartridge. The load usually represents the maximum charge of powder which: (1) fills up between 87% and 93% of the remaining case capacity (RCC) and (2) at the same time produces the maximum chamber pressure for the cartridge for which it may have been designed to handle.

On the other hand, bullet manufacturers tend to provide enormously comprehensive reloading manuals and handbooks to serve the entire shooting community with an exhaustive supply of technical data and recommendations. Whether a sheet of paper or a large book of data, we recommend that each shooter and reloader collects and keeps everything he finds on the subject.

1.2.1 THE DIFFERENCES BETWEEN SMOKELESS GUNPOWDER AND BLACK POWDER EXPLOSIVES

If we were to place a very small portion of smokeless gunpowder (a progressive burning powder) and also a small portion of black powder (an explosive) onto a mental pan or plate, the difference between the two will become immediately obvious once we have ignited them with a spark or fire.

Ignition of black powder will cause an explosion. All the potential energy burns within a fraction of a second taking the characteristics of a violent explosion and gone in a flash. Using black powder as the propellant in a gun, depending on the exact composition and ratio of the three components to each other as well as to the manufacturing process, black powder burns or explodes faster than the bullet can leave the barrel. In fact, a black powder revolver, as an example, may burn out completely before the bullet even leaves the cylinder, or it may completely burn out half-way through the barrel. Timing is extremely difficult, and the exact manufacturing process is critical to the timing. Timing is controlled by the composition of black powder, which is still, an explosive!

Smokeless gunpowder, on the other hand, is a progressive burning propellant – not an explosive. When we ignite it with a spark or fire, the potential energy in the fuel will not burn out completely immediately. Instead, it burns faster and faster over a period of time until it runs out of fuel and then stops. Igniting several grains of smokeless gunpowder in a pan or plate causes the flame to rise higher and higher as it burns faster and faster until it runs out of fuel and then stops – just as abruptly.

It is very different than black powder or any other explosive. Again, timing is critically important. We want the powder to stop burning at the precise moment the bullet leaves the barrel. If it stops burning too soon or too late, it adversely affects accuracy downrange. In that respect, it is no different than black powder. The only real difference is our ability to control the burning rate characteristics and time-pressure curves of smokeless gunpowder for superior consistency of accuracy and also burns cleaner with much less remaining residue in the barrel or cylinder, if a resolver. Residue in the barrel, as it accumulates with each successive shot, adversely affects accuracy until it becomes almost impossible to obtain any degree of consistency and accuracy. Black powder requires we clean the barrel frequently or preferably after each shot, while we can easily shoot with smokeless gunpowder literally thousands of rounds without cleaning it. One way of looking at this subject of black powder vs. smokeless gunpowder is to compare the operation of an automatic weapon, such as the M-16, to the differences in the performance characteristics of black powder and smokeless gunpowder.

As we write this, there are well over 100 smokeless gunpowders available on the open market for the recreational shooter in the United States. They range from extremely fast, in their burning rate characteristics, to extremely slow. We have included in this text 38 smokeless gunpowders available for rifles; 25 for shotguns; and another 25 for handguns. Many other powders are available but not listed here. Many of them are interchangeable. Many are not. We can use some very fast powders such as Unique or SR-4756 for shotguns, handguns, and even rifles for special single-shot applications, particularly with cast lead bullets, and others suitable for automatic weapons such as H-335 or IMR-3031 (notably for the M-16) in mid-range between very fast and very slow. Or we can go into the very slow rifle powders such as IMR-4831 and H-570, for the .50 caliber machine gun or some of the semi-automatic .50 caliber sport rifles presently on the market. Each smokeless powder, from the very fast to the very slow, has its value and application – and none is useless.

Black powder has its value and application, too, although not as flexible and versatile as modern smokeless powders. It is also very messy and greasy. That can be true with smokeless powders as well, depending on the lead alloy, type, chamber pressure, and lubricate.

If we can safely load black powder into the cartridge case for the M-16, the 5.56 × 45 mm, we will find enough energy to propel the bullet out of the barrel; but the powder will burn so fast, being an explosive, it will not produce the proper time-pressure curve to operate the automatic action. Instead, the spent cartridge case will stay in the receiver, immediately in front of the bolt, unable to extract the spent cartridge case or to reload the chamber with a new unspent cartridge. Then, there will an enormous greasy mess in the chamber and barrel making it necessary to clean it, or it may not even be possible to place a new cartridge into the chamber (Figure 1.1).

We can easily obtain the same results with very fast or very slow smokeless powders as well until we find the right powder to produce the

FIGURE 1.1 The M-16A1.

correct chamber pressure and time-pressure curve to propel the bullet out of the barrel, for the powder to stop burning at the precise moment the bullet leaves the barrel and, at the same time, the automatic action extracts the spent cartridge case and chambers a new one ready to fire.

The powders below are *only* approximate in relation to each other. Each powder burning rate characteristics will shift in relation to each other with a change in temperature to its relationship to the remaining case capacity, bullet weight, type and strength of crimp around the cartridge case, cartridge dimensional characteristics including thickness of the cartridge case walls, depth of bullet in the case, bullet distance to the throat of the bore, and certainly to the dimensional characteristics of the bore (including bore diameter) and length of the barrel.

Additionally, working out this chart at first became highly complex and difficult but later nearly impossible. When we wrote a letter to each gunpowder manufacturer for technical assistance, we discovered the existence of great reluctance to cooperate with us. With one exception, each of these manufacturers had either ignored us or had responded with a direct refusal. One manufacturer, upon receipt of our letter, responded through their attorney who promptly wrote a letter in an obvious attempt to intimidate us to stop asking for technical assistance. So, we sincerely apologize for any inherent inaccuracy in this comparison chart. We could do no better under the circumstances.

As we stated above in the image of the M-16A1, black powder burns all of its energy in one big explosion instead of gradually burning up its fuel and building up pressure over a period of time as is true with smokeless powder. This gradual build-up of pressure over time, the time-pressure curve, is necessary – not just to operate the automatic action of an automatic weapon such as the M-16A1 – but to time it to stop burning at the precise moment the bullet leaves the barrel. That is not possible or easy to do with black powder.

We need a *progressive burning* powder that burns up all its fuel more gradually over a longer period of time than is possible with an explosive such as black powder. When we look at Table 1.1, for the rifle powders in the comparison chart for gunpowders today, we will see 38 different powders suitable for a rifle. Only a small number of them, however, will work suitably for the M-16A1 and other automatic weapons. To work successfully, they must produce a chamber pressure for which the cartridge was designed to handle (50,000 to 55,000 psi (pounds per square inch)) and, at the same time, to burn its fuel slowly enough to stop burning when the bullet leaves the barrel and to allow the automatic action to extract the spent cartridge case and to chamber a new round.

Almost all the rifle powders will generate enough energy to produce the correct chamber pressures common with today's modern rifle cartridges, but only a few of them will provide the correct time-pressure curves to work an automatic action.

As an experiment, we can start off reloading five cartridges for the M-16A1 with the fastest rifle powder by DuPont Chemical Company, the IMR-4227, and to shoot them for a group downrange at 100 yards. IMR means "Improved Military Rifle" powder.

Then, we can reload another five cartridges but, this time, for a slower powder, the IMR-4198. Fire all five rounds at a target downrange at 100 yards, measure the group, and document it.

Reload another five cartridges with even a slower powder, this time the IMR-3031. Do the same thing as with the last two loads.

Continue reloading five more cartridges with each the IMR-4894, IMR-4064, IMR-4320, IMR-4350, and IMR-4831 powders.

A pattern will emerge almost immediately. Starting off with a load of IMR-4227, as the firing pin strikes the primer, the gunpowder burns rapidly to produce a chamber pressure of between 50,000 and 55,000 psi; the bullet goes through the barrel, but the powder burns so fast it stops burning before the bullet leaves the barrel. Because the time-pressure curve is far too short for the requirement of the action to work correctly, in spite of

TABLE 1.1 A Comparison Chart for Smokeless Gunpowders

Shotgun powders	Rifle powders	Pistol powders
(Fast)	(Fast)	(Fast)
Top Mark	Unique	Bullseye
450-LS	SR-4756	N-1010
Gray-B	2400	Hi-Skor
AA-125	H-110	PB
Red Dot	IMR-4227	230-P
N-2010	680	231
700-X	N-200	Red Dot
160	RX-7	Top Mark
AA-205	103	630-P
Green Dot	H-427	631
PB	IMR-4198	Unique
Unique	748	SR-7625
162	N-204	HS-5
SR-7625	102	AL-5
HS-5	IMR-3031	SR-4756
N-2020	RX-11	HS-6
500-HS	H-335	AL-8
AL-7	H-375	H-240
Herco	H-BL-C (2)	Herco
164	N-201	N-1020
SR-4756	H-4895	N-110
HS-6	N-203	2400
540-MS	IMR-4895	H-4227
AL-8	IMR-4064	IMR-4227
(Slow)	IMR-4320	IMR-4198
	101	(Slow)
	760	
	RX-21	
	H-414	
	IMR-4350	
	780	
	N-204	

TABLE 1.1 (Continued)

Shotgun powders	Rifle powders	Pistol powders
	100	
	H-450	
	H-4831	
	IMR-4831	
	N-205	
	H-570	
	(Slow)	

the proper chamber pressure, it will fail to complete its cycle of extracting the spent case or re-chambering a new one. Most likely, the bolt will not even move more than a half an inch or so, making it necessary to manually remove the spent cartridge from the receiver.

With a load of IMR-4198, the bolt will move a little further than with the IMR-4227, being a little slower in its burning rate characteristics; the cartridge case may move approximately halfway through the port before jamming the action. It will not leave the receiver, nor, again, will the action complete its full cycle of extraction of the spent cartridge case or re-chambering a new round. Groups downrange will be a little smaller than with the IMR-4227 but, still, unsatisfactory.

This time, however, with a load of IMR-3031, the burning rate characteristics will correspond precisely with the correct time-pressure curve to allow the action to operate a full cycle of extraction of the spent cartridge case and to re-chamber a new round. Groups downrange will be the smallest of the three loads and typically less than one inch in diameter at 100 yards. With 25.5 grains of IMR-3031 with a 52 grain spitzer boat-tail bullet, the muzzle velocity at approximately between 3,100 and 3.250 fps (feet per second), the powder, bullet, cartridge case length and neck diameter and thickness, chamber pressure, and time-pressure curve will be precisely correct for the need of the action to operate correctly (Figures 1.2 and 1.3).

As we continue to advance beyond IMR-3031 to IMR-4895, upon firing the round, the bolt moves much more slowly and may travel backwards about two-thirds from the breach before it stops, leaving the spent cartridge case in the receiver with the automatic action unable to move

FIGURE 1.2 A typical appearance of black powder.

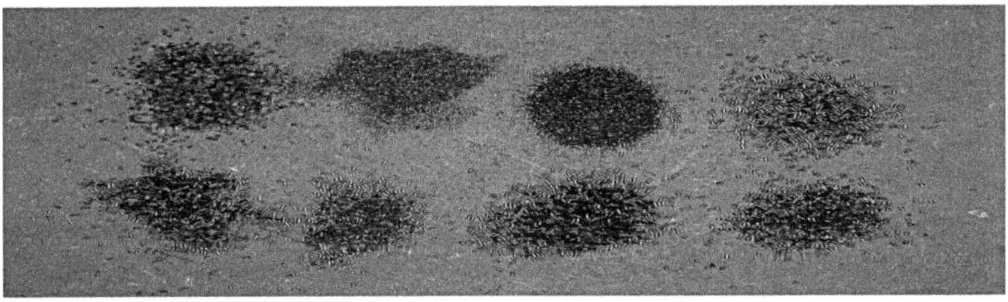

FIGURE 1.3 Extruded gun-powder into cylinders similar to the DuPont's IMR series is shown above with different shapes, diameters, and lengths to control the burning rate characteristics and time-pressure curves with additive mixtures to reduce flash, flame temperatures, smoke, leftover residue in the receiver and the barrel and sensitivity to the change in ambient temperature. Gunpowders can also be small spherical balls or flakes, depending on the manufacturer and application.

any further. The burning rate characteristics and time-pressure curve are no longer in time for the action to work correctly.

As we progress to IMR-4064, IMR-4320, IMR-4350 and, finally, to IMR-4831, we will find each successive powder to burn more slowly causing the action to work more sluggishly until it fails to move the action entirely, an event when no longer possible to stuff enough powder into the cartridge case to produce enough chamber pressure to operate the action.

As we progress from the fastest powder IMR-4227 to the slowest powder IMR-4831, the pattern is easy to perceive and recognize. Although IMR-4227 can produce enough chamber pressure, the time-pressure curve (burning too fast) is simply too fast to allow the operation of the action to complete its full cycle of retraction, ejection, and re-chamber of a new round. When we approach the correct powder, however, a powder capable of producing the correct chamber pressure and time-pressure curve for that action, the IMR-3031 (or H-335 and others), the action is fast, efficient, and smooth. As we get away from that correct powder, to slower powders, the action becomes more and more jerky and sluggish until there is no movement whatsoever.

Again, as we approach the correct powder (or powders), the groups downrange gets smaller and smaller until it reaches the design limitations of the gun system and, as we leave the correct powder, the groups downrange gets bigger and bigger.

Our conclusion is obvious: to be successful, for a given gun system, whether a revolver, rifle, shotgun, or automatic, we have to load ammunition with the correct powder that will produce the correct chamber pressure for the cartridge with the correct time-pressure curve for optimum results. Of course, we have other requirements as well. For consistency, from shot-to-shot, every cartridge case must be as identical to each other as technically possible in diameter, thickness, and length (particularly the thickness and length of the neck), the base of the bullet as round as technically possible, the powder in each cartridge case as identical in its chemical makeup as technically possible, the primers for each case as uniform as the manufacturing process allows, the correct rate of twist in the barrel for the bullet diameter and velocity and, most importantly, the quality of the firearm itself. It must be first class.

Figure 1.4 is a typical characteristic curve to show the relationship between time and pressure (the time-pressure curve). "X" represents the point where the bullet leaves the barrel at the precise moment the powder stops burning (see Figure 1.5).

1.3 TIMING IS EVERYTHING

It is important for us to "time" the time-pressure curve (Figure 1.4) and burning rate characteristics of the gunpowder to correspond with the

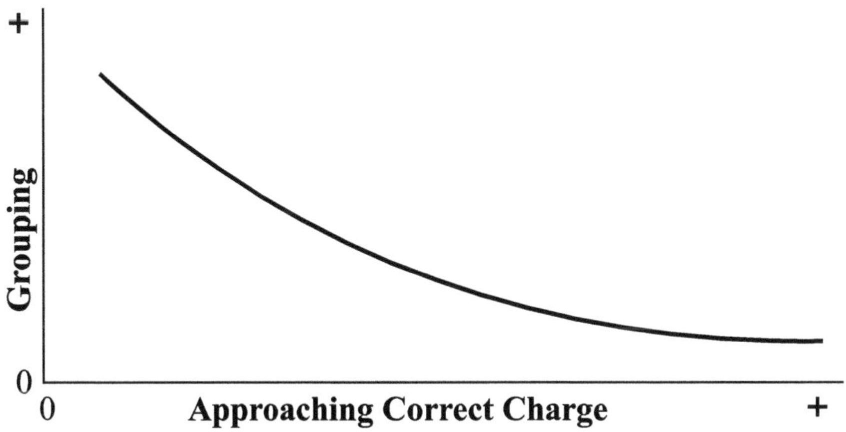

FIGURE 1.4 The time-pressure curve.

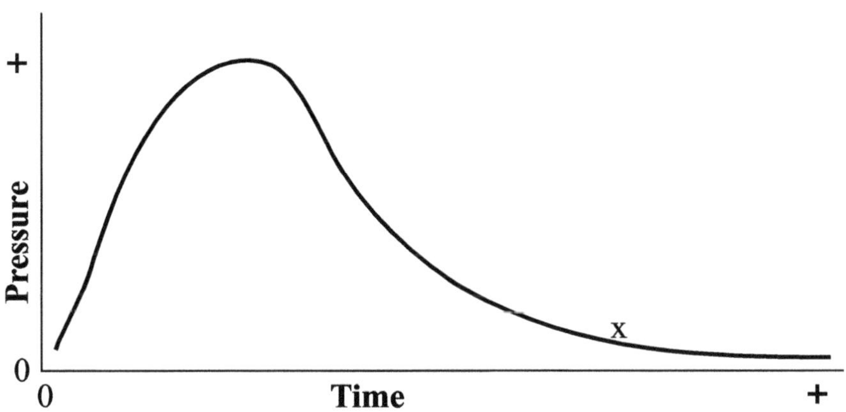

FIGURE 1.5 "X" represents the point where the bullet leaves the barrel at the precise moment the powder stops burning.

length of the barrel and velocity of the bullet. At the precise moment the bullet leaves the barrel, the powder ought to have finished burning (X). If the powder continues to burn after the bullet leaves the barrel, however, or stops burning before it leaves the barrel, we will have experienced a noticeable loss of efficiency with a corresponding increase in the size of the groups downrange.

We can note, when shooting in the wintertime with snow on the ground, if the powder continues to burn after the bullet leaves the barrel, we will find a streak of unburned powder on the snow extending several feet beyond and in front of the barrel. If the bullet is of a lead or a lead alloy, with lubricant, we may also see a streak of unburned or unused lubricant on the snow as well.

The characteristic curve in Figure 1.5 easily tells us; as we get closer and closer to the correct powder charge for a given barrel and bullet, the groups get progressively smaller.

1.4 THE CORRECT RATIO TO THE REMAINING CASE CAPACITY

Experience with the ammunition industry and recreational shooters has taught us the importance of filling up most of the remaining case capacity (RCC) (the distance between the primer and the base of the bullet in the cartridge case) with gunpowder in order to obtain the correct chamber pressure the cartridge was designed to handle for the maximum velocity and operating efficiency. If it is a fast powder that does not fill up the remaining case capacity for at least 87% to 90%, then the common practice among reloaders in the recreational community is to place fillers in between the powder and the base of the bullet. That filler keeps the powder from moving around in the case to provide a more uniform ignition from shot-to-shot and a more consistent group downrange.

Ammunition manufactures, on the other hand, tend to use the powder that will fill up most of the remaining case capacity for a given cartridge and bullet and, at the same time, consistently produce the correct chamber pressure the cartridge and gun was designed to handle.

Experiments have determined that moving the powder around inside of the cartridge case, when moving the gun, will also affect the burning rate characteristics and time-pressure curves from shot-to-shot, and that will affect accuracy downrange.

Experienced reloaders as well as the ammunition industry have learned to use the correct powder for a given cartridge and bullet combination with the correct ratio between powder and the remaining case capacity, to ensure consistency in chamber pressure, the burning rate characteristics and time-pressure curve. Without uniformity and consistency, from load to load, consistent accuracy is impossible (see Figure 1.6).

In Figure 1.7, we can easily see the effect of a constant ratio versus a variable ratio of powder charge to the remaining case capacity.

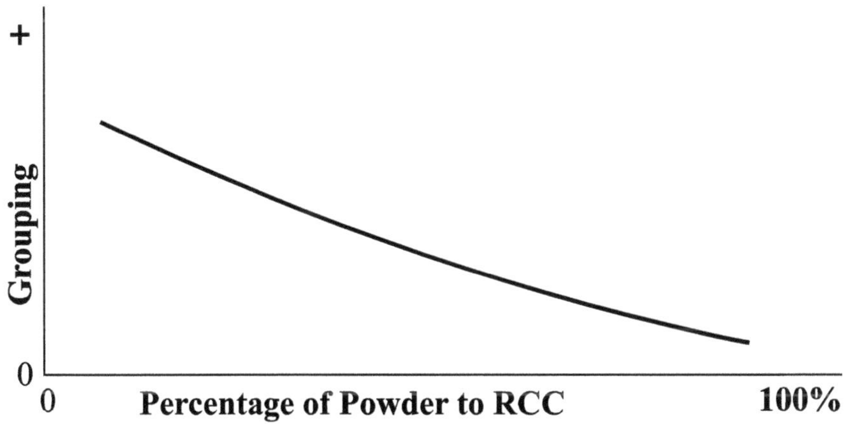

FIGURE 1.6 Consistent accuracy from load to load.

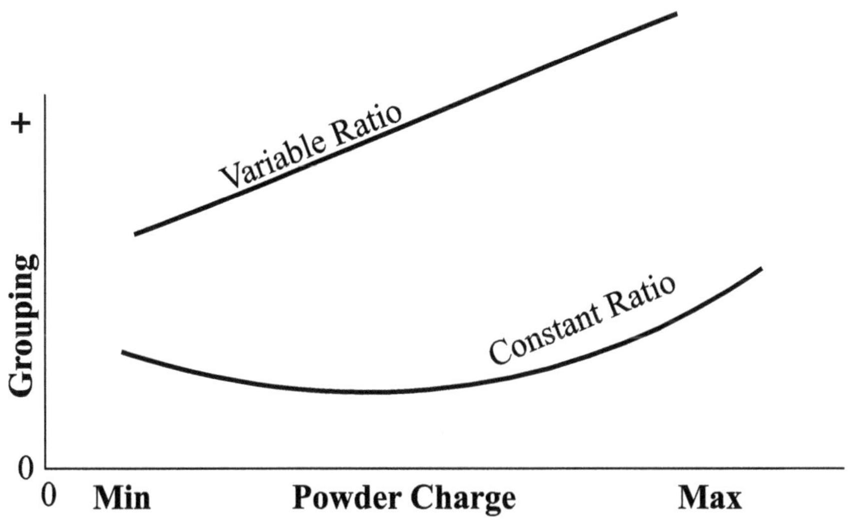

FIGURE 1.7 The effect of a constant ratio versus a variable ratio of powder charge to the remaining case capacity.

It is impossible to maintain accuracy with a constantly changing ratio of powder charge to the remaining case capacity. Then, in Figure 1.8, we can see, as we approach the correct powder charge ratio for the remaining case capacity, the groups' downrange gets smaller and, as we deviate from this correct ratio, the groups get bigger.

Likewise, starting from the fastest powder available and working toward the slowest, for a given remaining case capacity and bullet weight,

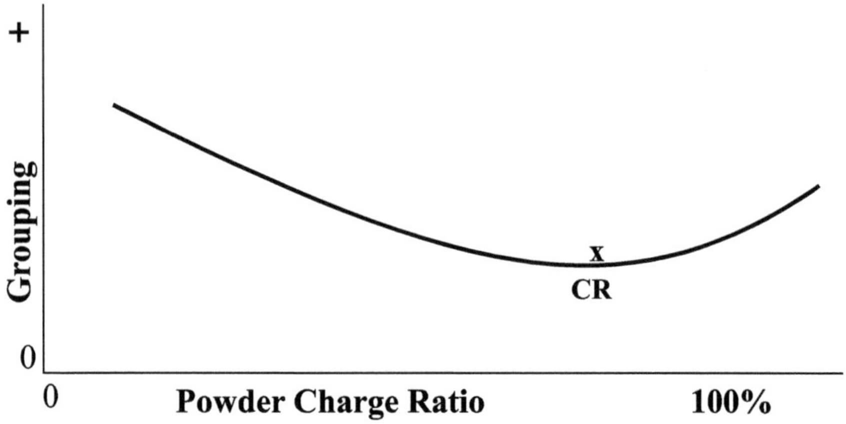

FIGURE 1.8 The correct powder charge ratio for the remaining case capacity.

in Figure 1.9, the groups will progressively grow smaller as we approach the correct powder (CP).

However, as we go beyond the CP, the groups will again progressively grow larger until it becomes almost impossible to hit anything with reliability and consistency (Figure 1.10).

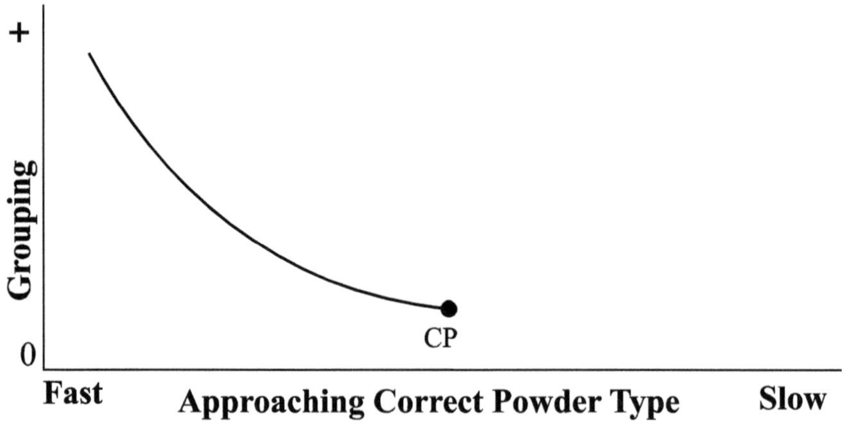

FIGURE 1.9 The fastest powder available and working toward the slowest, for a given remaining case capacity and bullet weight.

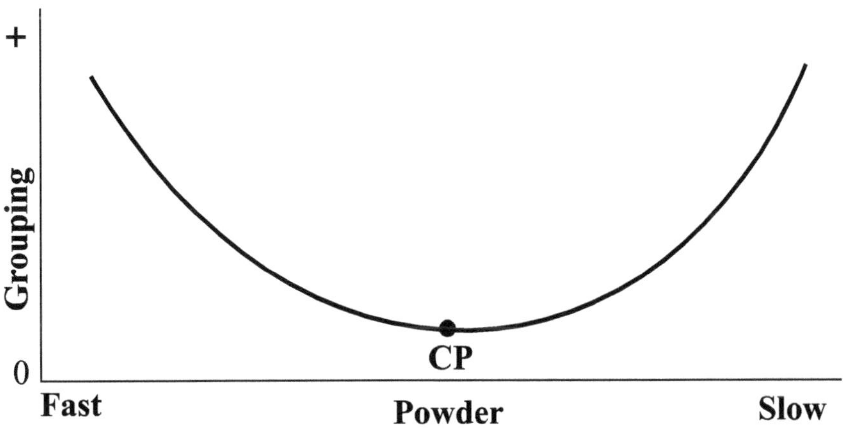

FIGURE 1.10 The groups progressively grow larger until it becomes almost impossible to hit anything with reliability and consistency.

Then, again, this same pattern holds true with automatic actions such as the M-16 and all its derivatives (Figure 1.11), and every other automatic or semi-automatic weapon. As we move from the fastest powder to the slowest powder, the speed of the automatic actions vary with the speed of the powder.

As we start loading cartridges with the fastest powder, we will soon learn that the powder burns so fast that the automatic actions will not operate; then, as we use a slower and slower powder, the actions begin to move and finally to operate properly. As we approach the CP for a given action,

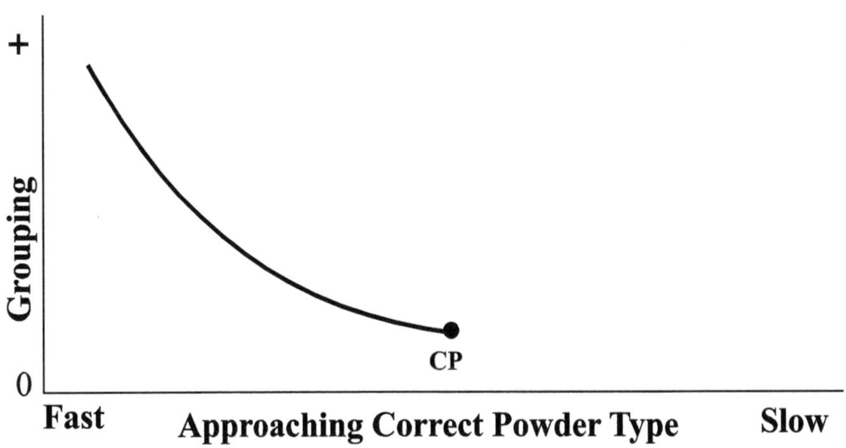

FIGURE 1.11 The same pattern of Figure 1.10 with automatic actions.

the action runs smoother and smoother with less and less jerkiness. When we reach the CP, the action operates perfectly: fires, extracts, ejects, and chambers a new cartridge in the correct sequence.

As we go beyond the CP with progressively slower powders, the action begins to operate more and more sluggishly until it fails to operate entirely, an event that occurs when there is not enough powder to produce the correct chamber pressure (Figure 1.12).

On the other hand, as we can see in Figure 1.13, as we move from the fastest powder to the CP, to operate the action perfectly, accuracy progressively improves until it corresponds with the CP.

When we go beyond that point, accuracy progressively decreases with the increase in the sluggishness of the action.

1.5 RELOADING AMMUNITION FOR A PARTICULAR GUN

While we must maintain the CP and ratio of powder to the remaining case capacity, in order to obtain the maximum accuracy potential, we will not succeed unless we also stabilize the bullet with the correct muzzle velocity and bullet relative to the rate of twist.

Two methods are available: (1) we can manipulate the powder charge to control the muzzle velocity or (2) we can vary the weight of the bullet to match the rate of twist (see Theory of Twist).

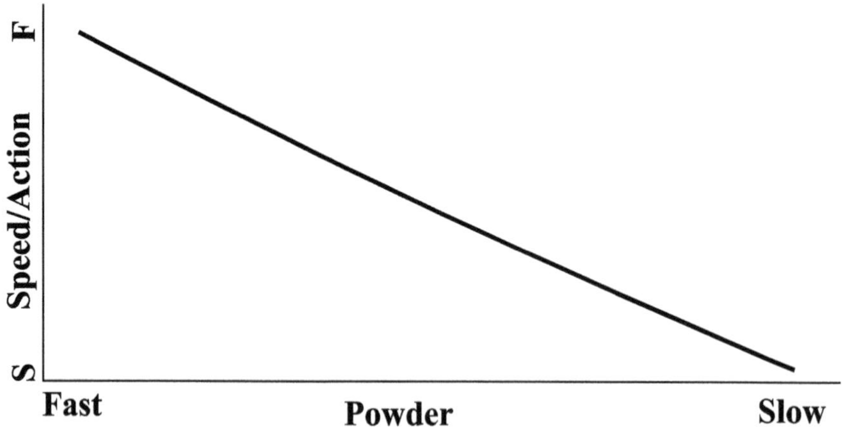

FIGURE 1.12 The approach of the correct powder results in the action to runs smoother and smoother with less and less jerkiness.

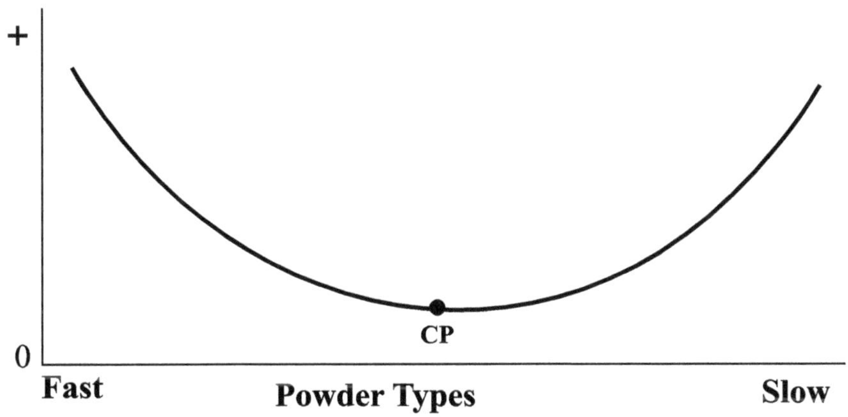

FIGURE 1.13 An event that occurs when there is not enough powder to produce the correct chamber pressure.

To reload ammunition for a particular gun, whether handgun, rifle, shotgun, machinegun, or submachinegun, we must take the following decisions:

(1) application;
(2) remaining case capacity;
(3) bullet length; and
(4) the barrel's rate of twist.

Application is the intention of use. How do we intend to use the gun? For hunting, target practice, competition, self-defense? In close-quarters? At great distances? Big-game hunting? Small-game hunting? Dangerous game? Once we decide upon the application, we can easily select the muzzle velocity and bullet length (with the accompanying felt recoil and recoil jump).

Bullet weight determines the length of the bullet for a given diameter (caliber), which determines the remaining case capacity in a given cartridge case. The rate of twist has already been determined by the gun's manufacturer and, with some exceptions, almost never is an option for the buyer. It will almost always be wrong except for the original design application (such as big-game hunting), particularly for handguns (such as self-defense).

Because we can almost never have the option to vary the barrel's rate of twist in order to accommodate a particular cartridge and bullet length, weight, and muzzle velocity, we will have to manipulate muzzle velocity and bullet length, through its weight, to obtain the proper bullet spin for the maximum accuracy potential downrange.

Most guns, with the important exception of military weapons – which are always precisely designed for extreme reliability with a certain cartridge, bullet weight, gunpowder, and muzzle velocity – were designed many years earlier for heavier bullets using gunpowders that may not be available any longer. Many hunters prefer to select the bullet weight, for their particular gun, that will produce the greatest muzzle velocity, thinking of it as superior to heavier bullets that produce much less muzzle velocity. That automatically relegates them to a lighter bullet, sometimes the lightest bullet available on the market for any given cartridge and caliber. Because the rate of twist may have been originally designed for the heaviest bullet, or a much heavier bullet, there will always be a certain incompatibility with a built-in inaccuracy when using a different weight, something most hunters and recreational shooters will not perceive or understand until much later after greater experience. Lighter bullets, fired out of the proper cartridge case at the correct chamber pressure, will most definitely provide for a flatter trajectory but lose momentum and velocity much faster than a heavier bullet, even if the heavier bullet starts off at a substantially lower velocity.

With bullet manufacturers manufacturing bullets from light to heavy, for a given caliber and cartridge, the reloader has an enormous advantage over single users using ammunition from off the shelf at a local retailer. He can reload ammunition with extreme precision for extreme accuracy for his particular gun, ordinarily not an option for anyone else using off the shelf equipment.

When we place a given bullet in a cartridge case at its proper depth, sometimes something we do arbitrarily on the basis of our perceived physical attractiveness of the whole cartridge, it will automatically determine the amount of space between the primer and the base of the bullet to provide for the remaining case capacity. That will determine the kind and the amount of powder we can use to obtain the correct chamber pressure.

We have two methods of measuring the remaining case capacity:

(1) The first method to accurately measure the remaining case capacity is to put water into a spent cartridge case, with a spent primer still inside, at the depth corresponding with the exact position of the base of the bullet. Then, we weigh the water.
(2) The second method is to weigh the exact amount of powder we intend to use at the same depth. This method, if we use it correctly, with the CP, is much more accurate.

Once we know the remaining case capacity, we then calculate 87% of it to give us the CP charge. Some powders require a slightly different ratio of charge to the remaining case capacity; however, on the most part, 87% is a fair estimate at the proper ratio.

After we determine the remaining case capacity, with any chosen bullet weight, we must decide on the muzzle velocity we need to stabilize for the rate of twist on the gun we have chosen to use.

When we finally select the CP and bullet length/weight and muzzle velocity, it becomes imperative to test and then "tune" the ammunition for the gun.

1.6 TESTING AND "TUNING"

The following two methods work very well:

(1) This first method is the best. We load a 5-shot group, with the correct components, with the powder filling up 87% of the remaining case capacity; then working at even or uniform intervals (1/10, ½, 1, 2, 3, etc.), we load a series of 5-shot groups at these intervals working toward 5% below and 5% above the 87% RCC. Sometimes, it may become necessary to go 10% below and 10% above the 87% RCC, particularly with very slow burning powders in large-capacity cartridge cases.

    Through a pair of chronographic screens, we fire each 5-shot group to electronically measure the velocity of each round. Then we calculate the *average velocity,* the *medium velocity*, the *extreme spread* (difference between the lowest and the highest velocities), and the *standard deviation*, the most important indication of both accuracy and performance of ammunition and gun. The 5-shot group with the lowest standard deviation will be the most important clue to its accuracy, and hopefully somewhere between 2 and 6 fps for handguns with small remaining case capacity and fast powders and 20 and 30 fps for rifles with large remaining case capacities and much slower powders.

Of course, if the average velocity is nowhere near the velocity compatible for the rate of twist, we will have to either increase or decrease the powder charge or change the powder type entirely to something either slightly faster or slower in its burning rate characteristics. Usually, a low standard deviation will correspond with the correct muzzle velocity at the given rate of twist. It will also prove that each important variable responsible for accuracy and consistency is under control and the gun works correctly.

(2) In the second, less effective method to test and tune our ammunition for a given gun, instead of a pair of chronographic screens, which is very expensive for many people, we simply shoot at a paper target for the smallest groups; measure the size of each 5-shot group; calculate the *average velocity*, the *medium velocity*, the extreme spreads, and finally their standard deviations.

Essentially, the problem with this second method is the introduction of several variables, which is not a problem with the first method: Namely, it is the psychological anticipation of felt recoil (flinching), ambient temperature from the muzzle of the gun to the target downrange and a much less consistent sighting picture.

1.7 FURTHER REFINEMENT

Unless our ammunition performs with a standard deviation to our satisfaction, it will become necessary to "tune" it up for further refinement.

At this point, a whole array of possible problems and their accompanying solutions may pop up in full view of the serious shooter.

We can "tune" the cartridge neck for greater uniformity in roundness, thickness, and length. Every time we shoot a gun, the cartridge case expands and then contracts by perhaps 2,000th to 3,000th of an inch, sometimes less and sometimes more, depending on the interior dimensions of the chamber. It is always a little larger in diameter after we fire it and then that requires we *re-size* it before reloading it again. There are special re-sizing dies just for that purpose. Each cartridge case progressively enlarges in length until, eventually, it cracks from fatigue. This crack can occur anywhere on the cartridge case, from somewhere near the rim, in the middle of the case, or on the neck itself. If we fail to recognize this fatigue and subsequent crack, that particular round will destroy the group downrange.

With the proper tools, most of them readily available in gun shops and online, we can ream the interior of the neck for uniformity and then mill the exterior of the neck to ensure uniformity in both diameter and thickness. Then, with another tool, we can trim the length of the neck for uniformity in length. That is important. If the length of the cartridge case varies, from case to case, by as much as a few thousandths of an inch, it will undoubtedly adversely affect the groups downrange, particularly with handgun cases.

Just as importantly, the depth and diameter of the hole in the base of the cartridge case, to accommodate the primer, changes constantly with each shot. For uniformity, we must also ream this primer hole before we reload the cartridge for the next shot.

In addition, drilling the primer flash holes with an ordinary drill bit does the same thing by making each flash hole uniformly round in diameter. That tool is also readily available on the market, in gun shops, hardware stores, and, of course, online.

Trimming the cartridge cases, particularly handgun cartridge cases, always significantly improves accuracy. Back in the summer of 1976, we observed a 57% improvement in accuracy with .44 Magnum cartridges and a Smith & Wesson M-29 revolver. We consistently got groups small enough at 100 yards, in a standing double-hand hold, to hit (empty) beer cans without difficulty. Do not do it with real beer in the cans. Some of the shooters might get very angry with you.

Apparently, handgun cartridges with fast powders are much more sensitive to uneven lengths than rifle cartridges using much slower powders.

These improvements occur as a direct result of a greater uniformity in the time-pressure curve of the powder from cartridge to cartridge. When the powder burns more uniformly, shot after shot, we will always find a smaller variation in muzzle velocity and, therefore, a smaller standard deviation and tighter group.

In a revolver, as an example, it becomes important to make each chamber in the cylinder as uniform in size, roundness and true to the center of the bore as technologically possible.

Burrs in a barrel's rifling can make the time-pressure curve of each successive shot vary significantly as lead successively builds up and successively increases resistance to the bullet as it travels through the bore. Each successive shot will then adversely affect standard deviation and, of course, the accuracy downrange. Burrs and lead build-up on the sides of the lands of the rifling destroys the "timing" of the bullet as it travels through the bore causing the powder to stop burning before the bullet leaves the barrel. Remember! Proper timing requires the bullet to leave the barrel at the precise moment the powder stops burning. If it should stop burning before the bullet leaves the barrel, or continues to burn after the bullet leaves the barrel, or, if it should vary in this interior ballistic behavior, from shot to shot, or progressively deteriorates as the lead builds up in the rifling from burrs, with the bore getting dirtier and dirtier, except for very big targets at very small distances, hitting anything is only by chance. Shooting for groups is nearly impossible.

One of the most effective solutions to this common problem, when shooting lead alloy bullets, is to lap the bore with a good quality lapping compound. They are readily available in the market just about anywhere.

CHAPTER 2

A PRACTICAL APPLICATION TO SCIENTIFIC EXPERIMENTATION

CONTENTS

2.1 INTRODUCTION

There is in the total body of science a philosophy of science and a science of experimentation.

Though any acutely intelligent reader of this scientific text on ballistics can easily perceive its philosophy or, more appropriately, the set of attitudes guiding and governing the writer's methods of scientific research and database development, it is clearly beyond his scope to write from a philosophical viewpoint.

Nevertheless, we have an urgent need to elaborate on the practical methods of scientific experimentation for the recreational shooter who lacks the monetary resources to organize and maintain a comprehensive scientific laboratory. With that limitation in mind, we can still develop a more humble laboratory for the purpose of valid scientific research and development.

Let us start with a common problem each of us in the American shooting community can easily understand immediately and without any real effort.

2.2 IDENTIFICATION OF OUR BALLISTIC NEEDS

Suppose we were to purchase a new Smith & Wesson Model 29.44 Magnum revolver with a 4-inch barrel. After honing and stoning the trigger assembly for a smooth, crisp, and reliable action, we are now ready to examine the gun ballistically.

The barrel's rate of twist is a fast 1 in 18¾ inches. It means the bullet spins through the bore at a rate of one revolution every 18¾ inches.

A typical bullet for the .44 Magnum has an outside diameter of approximately 0.429 inches after going through a re-sizing die; a length of 0.75 inches and a weight of 240 grains (7,000 grains per pound).

If we were to refer to the section on the Theory of Twist, we will find an equation to calculate the best muzzle velocity (BMV) for this bullet and the proper rate of twist in order to properly stabilize this 240 grain bullet.

However, with this typical bullet and the gun's rate of twist, the BMV is extremely low at 835 fps. Obviously, we are either shooting the wrong bullet, at the wrong muzzle velocity – 1100 to 1300 fps commonly with off-the-shelf ammunition (depending on the length of the barrel and the powder type) – or the gun manufacturers use the wrong rate of twist in this caliber. Actually, with that particular bullet with that stated rate of twist, it would be more appropriate for the .44 Special cartridges instead of the .44 Magnum. Nevertheless, we can use that .44 Special load very effectively in this .44 Magnum revolver using either the .44 Special or .44 Magnum cartridges (as long as we use fillers). Without an exaggeration, such a load will allow sufficient accuracy, shot after shot, to consistently hit Budweiser beer cans at 100 yards with a standing, double-hand hold. Make sure they are empty Budweiser beer cans!

Referring back to the Theory of Twist, we can calculate the best bullet length (BBL) for a 1 in 18.75 inch rate of twist at – say, 1200 fps. But we will run into another problem. Even at a lower muzzle velocity of 1150 fps, the correct bullet length becomes 1.03 inches with a weight in excess of 300 grains. If we were to drop down to 1100 fps, we can keep the bullet's length to less than 1 inch, or 0.988 inches, to fit the cartridge into the chamber and cylinder. Even so, we will continue to have serious problems. Gunpowders commercially available to us for this caliber and the remaining case capacity were not designed to handle a bullet weight in excess of 300 grains at this range of 1100-1200 fps in a revolver. True, we can achieve such velocities for compatibility to the rate of twist; however, felt-recoil would be most unpleasant and only suitable for serious big-game hunting or long-distance silhouette shooting. Again, as long as we are willing to accept the punishing recoil and can avoid flinching, accuracy will allow us to hit the same (empty) Budweiser beer cans as with the lighter loads and lighter bullets, and if we can re-design the grip to fit our hand more ergonomically to human-factor engineer a reduction in felt-recoil.

Threshold of pain from felt-recoil is around 10 ft/lbs of kinetic energy with 4 to 8 ft/lbs quite pleasant, as with the M-16, and anything above 30 ft/lb – very painful. As a reference: a .50 caliber machinegun can easily produce in excess of 90 ft/lbs of felt-recoil and a .30/06 more than 30 ft/lbs.

Use of these slow-burning powders will prove burdensome, too. In some instances, we could never stuff enough powder into the cartridge cases to obtain the correct chamber pressure and muzzle velocity. In other instances, some fast-burning powders can prove dangerous to use. Finding a compromise is the real challenge. Remember! We must find a powder to fill up between 87% and 93% of the remaining case capacity – and, at the same time – produce the proper chamber pressure and muzzle velocity to stabilize with the rate of twist and burn with the correct timing.

The time-pressure curve must correspond with the velocity of the bullet traveling through the bore. As the bullet leaves the barrel, the powder must stop burning. If not, we will not obtain the maximum ballistic efficiency and accuracy downrange and, of course, within the design limitations of the gun. Just as importantly, if not more so, if the time-pressure curve varies with the velocity of the bullet traveling through the bore to exit the barrel, we will find big groups downrange, extreme velocity spreads, and large standard deviations (SDs). Accuracy is a lost cause. So, what is the solution?

2.3 SEEKING OUT A SOLUTION

Basically, our solution is one of two extremes:

(1) We can go to the big bullet at the high velocities and fool around with the problems of proper gunpowder selection and "timing," or
(2) We can go to the opposite extreme of selecting a fast-burning powder with the use of a standard 240 grain bullet and a filler to ballistically simulate the effects and characteristics of a slower powder.

Both solutions have their advantages and, of course, disadvantages. If we were to use the 240-grain bullet, we must stick with a muzzle velocity of around 835 fps to stabilize it properly, which makes the load into a short-range target and plinking round (and an excellent round for self-defense).

If we were to aim for the 300 grain bullet, at a proper muzzle velocity to stabilize with that given rate of twist, we must expect and accept the prospects of very uncomfortable recoil and recoil jump. Flinching would be very difficult to avoid for most of us.

However, the heavier bullet will have the advantage of extreme accuracy out to 200–250 yards for big-game hunting (if we don't flinch) and the lighter bullet a maximum effective range of between 150 and 175 yards.

Groups fall apart at 200 yards for the lighter bullet and at 275 yards for the heavier bullet.

Let us take a moment to make it clear this little dissertation or study on the .44 Magnum revolvers and the two different bullet weights, with their two extremes, represents the same patterns and relationships with every gun regardless of the caliber or the cartridge. We will always have the same problems of designing a load for every gun, cartridge, bullet, and rate of twist.

2.4 OPTIONS

Suppose we decide to stay with the typical and readily available 240-grain bullet. What are our options concerning powders? Actually, every powder on the commercial market today will represent some kind of problem with this 240-grain bullet at 835 fps, as with any other cartridge. Even the slowest-burning pistol powders would burn too fast for our purpose. Even use of some of the fast-burning rifle powders would give us ignition problems, and occasionally, bullets may lodge inside the bore when the powder fails to ignite and burn entirely. These rifle powders were clearly not designed for short barrel pistols with pistol primers and small remaining case capacities. They were designed for small capacity rifle cartridges with rifle primers.

Fast-burning pistol powders would, as we would expect, easily give us the correct muzzle velocity but will burn too quickly and fill up less than one-half of the remaining case capacity for the proper chamber pressure. They will also burn out of time. While some powders will burn entirely before the bullet leaves the cylinder, in the case of a revolver, others may complete the full burning cycle with the bullet halfway through the bore.

Then again, with fast-burning powders, we can use fillers to alter the time-pressure curve, burning rate characteristics, and hence the "timing" to make them behave as if they were slow-burning powders. If we do it correctly, the fast-burning powders – instead of at the end of the cylinder or in the middle of the bore – will stop burning at the end of the barrel, the way it should be.

To obtain the correct combination of powder and filler, we will need to experiment with a safe powder, such as Red Dot, basically a shotgun powder mid-range in its burning characteristics between the fastest pistol powder, Bullseye, and HS-5, a powder in between Bullseye and IMR-4198, the slowest pistol powder (see Figure 2.1). So, Red Dot is one quarter up

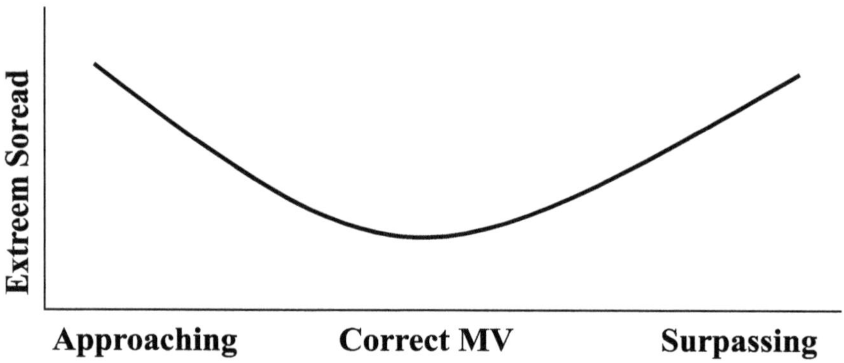

FIGURE 2.1

the scale. With the proper filler, we can make Red Dot behave as if it were halfway or more up the scale, as is true with any other powder with the right filler.

We can use anything from toilet paper to Oatmeal to Corn Cob as fillers. Certain fillers are obviously better than others.

Oatmeal works well as long as we do not have to meter it through a powder measuring device, or in a progressive reloading press, due to its irregular shapes and sizes, everything from extreme big flakes to extreme small flakes.

Corn meal actually noticeably increases the muzzle velocity and felt-recoil but, unfortunately, meters poorly. It tends to adhere to the metering machinery and frequently clogs the drop tubes to produce extreme variations from cartridge to cartridge.

Corn cob, due to its irregular shapes and sizes, will not meter well, either, unlike most ball powders, but meters much better than most other fillers. When it burns, it produces a unique odor of sweet cider; possibly, if someone were to do enough shooting with it, he could very well get a little "high" if the wind blows it back in his face every time he fires a shot. Corn cob softens the felt-recoil and dampens the report. Corn meal stiffens it and amplifies the report.

On the other hand, other fillers have no apparent effect over felt-recoil but always improve the ballistic efficiency of the ammunition, meaning a more consistent accuracy downrange from shot to shot. They seem to work by controlling the uniformity of the powder's burning rate characteristics and time-pressure curve. We also have commercially available fillers that do very well.

2.5 COMMENCEMENT OF AN EXPERIMENT

To start this experiment, we need to identify the exact amount of Red Dot it takes to produce a muzzle velocity of 835 fps without fillers.
On this particular day of July 4, 1986, 5.2 grains of Red Dot with a typical 240 grain bullet produced an average muzzle velocity of 841 fps, a figure very close to the 835 fps we require. Actually, as long as we are within plus or minus 50 fps, we are well within the "ball park."

Now, at this moment in time, we must stop long enough to explain the use of certain statistical data points, particularly SD. Some of our members of the shooting community and Academia, knowledgeable in the science of statistics, may question the validity of using SD in a distribution of five scores (five shots with five different velocity readings). They may retort that such a small distribution of scores in a statistical analysis to determine the central tendency (average muzzle velocity) of a distribution of scores and its SD has no real statistical value. In that context, they are absolutely correct. However, our purpose in the use of these statistical tools is not to establish a *definitive* analysis of the central tendency and SD, but to *identify* the patterns and relationships between a certain amount of powder to muzzle velocity to the bullet with a given the rate of twist.

Though certainly definitive, the central tendency and SD of – say, 100 or even 1,000 shots – would have very, very little value to the real world of shooting outdoors with a constantly changing environmental condition (see, The Field-Effect Theory and the Effect of Field-Effect Over Trajectory).

It would be a hopeless cause to expect anything to the contrary. Unless we intend to work out the mathematical relationships to a higher level or to establish an engineering database for engineering applications, we will waste a lot of ammunition. Except for mistakes in testing procedures, the patterns and relationships will stay the same, whether in a distribution of five scores or 1,000 scores. Only the data points change to either increase or decrease accuracy, or sometimes, to clarify or bring out patterns and relationships a small distribution of scores will not obtain ordinarily. In this context, we do not need or desire such extreme accuracy but only to identify the patterns and relationships that will allow us to "tune" up a load for a particular gun with a particular rate of twist.

2.6 PATTERNS AND RELATIONSHIPS

When we study the results of Groups I–VI, patterns and relationships become readily apparent (Table 2.1). As we approach the correct muzzle velocity for the rate of twist, ballistic efficiency increases.

TABLE 2.1

Group I	Group IV
4.8	5.1
1. 552 Lo=551	1. 743 Lo=743
2. 551 Hi=589	2. 758 Hi=758
3. 565 Av=565	3. 756 Av=750
4. 589 ES=38	4. 751 ES=15
5. 557 SD=16	5. 756 SD=10
Group II	**Group V**
4.9	5.2
1. 562 Lo=561	1. 835 Lo=819
2. 564 Hi=589	2. 855 Hi=855
3. 561 Av=570	3. 855 Av=841
4. 589 ES=28	4. 855 ES=36
5. 570 Sd=15	5. 819 Sd=17
Group III	**Group VI**
5.0	5.3
1. 652 Lo=635	1. 917 Lo=895
2. 651 Hi=663	2. 916 Hi=935
3. 662 Av=652	3. 935 Av=919
4. 663 ES=28	4. 933 ES=40
5. 635 Sd=11	5. 895 Sd=16

In Group I, the extreme spread was 38 fps; however, as we approach the correct muzzle velocity, the extreme spread successively decreases and then successively increases when we surpass the correct muzzle velocity.

Likewise, as we approach the correct muzzle velocity, SD successively decreases and then successively increases when we surpass the correct muzzle velocity.

All cases in these groups were un-crimped; though not definitive, there is another pattern. In five of the six groups (87.5%), the last shot is low, and in three of the six groups (50%), the last shot represents the lowest velocity in the group. Without further statistical analysis, what does these data suggest? (Figure 2.2).

It may suggest, without our crimping the cases, each shot in the revolver causes the bullet in each cartridge to move forward incrementally. Because the last cartridge is the last shot, each preceding shot may have caused a cumulative effect of the bullet protruding more than the others. Our experience will soon teach us, even without the assistance of an electronic chronograph, when we fire heavy or relatively heavy loads in a revolver without first crimping the cases, as the bullets move incrementally forward due to recoil from the preceding shots, the subsequent recoil softens and the report takes a successively softer sharpness with a successively lower range of audio frequencies and set of harmonics. Experience will teach us that this is also true with rifles and shotguns (excluding single-shot guns, of course). These statistical data also suggest a successive decrease

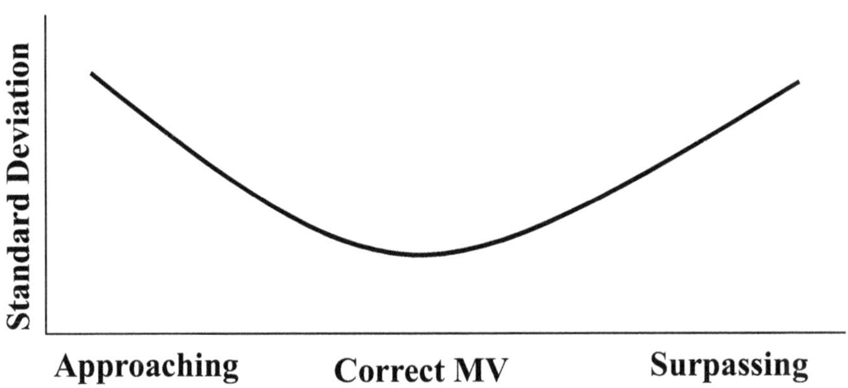

FIGURE 2.2

in muzzle velocity, which certainly corresponds with the above experiences. Recoil corresponds with muzzle velocity.

Then, most interestingly, the highest velocity is not the first shot, as we would expect logically, but either the third or fourth shot with one exception. In Group IV, the highest velocity was the second shot. Why? We would need to perform a series of exhaustive experiments to determine the cause of this interesting phenomenon. When we do not crimp the cases, it seems logical to assume all subsequent shots to be a lower velocity then the preceding shots, with the first shot the highest velocity and the last shot the lowest velocity. On the basis of our experience, we have discovered that, in spite of the high precision in the manufacture of cartridge cases throughout the world, the dimensional characteristics – length, inside and outside diameter, thickness, and metallurgy – always slightly vary from cartridge case to cartridge case. Even with a difference of 2,000th of an inch in any of the dimensions, it affects the performance of the round. The variation in the crimp always affects this performance, too and, when there is a metallurgical flaw in the cartridge case, such as when the case cracks upon firing it for the first time, the end results proves catastrophic. That proves very dramatic when we fire a group of five rounds and the crack causes that particular round to significantly head off from the center of the group downrange.

Also, as we shoot these loads at targets downrange, without first crimping them, we will find a pattern of the groups enlarging with each successive shot. When we properly and uniformly crimp the cases, there is no such pattern.

2.7 EXAMPLE

In early September of that same year, using the same bullet, primer, cartridge case, and lubricant, we re-tested Group VI with 5.3 grains of Red Dot. This time, though, the temperature was nearly 30°F lower than the preceding one on July 4, 1986. Cases were crimped.

Notice the difference in average muzzle velocity of 188 fps between the two groups. The only significant change in this test was the ambient

TABLE 2.2

Group VI (re-test)
Early September
1. 742 Lo=701
2. 730 Hi=752
3. 732 Av=731
4. 701 ES= 51
5. 741. Sd = 19

temperature. Otherwise, the loads were identical. Notice, also, there is a different pattern than the original Group VI (Table 2.2). The first shot is the highest velocity and the last shot next to the highest. We no longer have the same pattern of the muzzle velocity successively dropping as the bullet moves out of its case with each successive shot. With the exception of the fourth shot, each shot is about the same as the others with no distinctive pattern popping into view. We only need to improve the quality of the ammunition to drop the extreme spread and SD for better accuracy. Certainly, if we were to perform a more definitive statistical analysis of this group, we might very well find new patterns and relationships. Who knows?

Now, to improve the ballistic efficiency of the powder, what would happen if we were to use a corn cob filler to fill up the remaining case capacity following the gunpowder? With 5.3 grains of Red Dot in this cartridge case, there is substantially less than 87% of the remaining case capacity. We will need the filler to improve efficiency to keep the powder in one place of the cartridge case between the primer and the base of the bullet.

These tests in Table 2.3 were performed late in the afternoon after a significant drop in temperature, which may account for the insignificant increase in muzzle velocity relative to the re-test of Group VI. As a rule, corn cob filler will increase muzzle velocity by an average of 50 fps.

TABLE 2.3

Group VII	Group IX
5.3	5.3
1. 766 Lo=738	1. 769 Lo=729
2. 767 Hi=767	2. 749 Hi=769
3. 761 Av=756	3. 729 Av=755
4. 750 ES= 29	4. 766 ES= 40
5. 738 Sd= 12	5. 764 Sd= 16
Group VIII	**Group X**
5.3	5.3
1. 736 Lo=736	1. 730 Lo=730
2. 770 Hi=770	2. 746 Hi=807
3. 743 Av=751	3. 747 Av=755
4. 740 ES= 34	4. 746 ES= 77
5. 767 Sd= 16	5. 807 Sd= 29

Ordinarily, as we can see, Group VIII through Group X is extremely good. Average muzzle velocities vary from a low of 751 fps to a high of 756 fps – an extreme spread of 5 fps. When we shot these same loads at 150 yards, we got a group of about the diameter of a man's face (shooting in a standing, double-hand hold). At 100 yards, we were hitting bowling pins almost consistently (using the same hold). Then, at 200 yards, we were unable to hit a 12-foot high bank consistently.

Even though the medium muzzle velocity was 82 fps away from our objective of 835 fps, accuracy was extremely good. Perhaps, if we were to add another 1/10 of a grain of Red Dot to 5.4 grains, we might obtain this objective.

Incidentally, a magnum pistol primer will increase muzzle velocity by approximately 10 fps with this load, and possibly even more with slow-burning rifle powders. Changing the filler from corn cob to corn meal will increase muzzle velocity by another 50 fps, roughly.

Increasing the barrel's length will produce an average increase in muzzle velocity of about another 10–30 fps for every inch. It depends on the powder, primer, and filler as well as the ambient temperature.

Slower burning powders will increase muzzle velocity more dramatically with each incremental increase of barrel length than the fastest powders, which will usually stop burning long before the bullet leaves the barrel when we go beyond the correct barrel length.

Two or three grains of black powder or Pyrodex in front of the primer and in between the primer and a slower burning pistol powder, as well as a slow-burning rifle powder in a handgun cartridge, or large-capacity rifle cartridges with very slow-burning rifle powders, may have the same effect as a magnum primer. It works by increasing the ignition temperature and shortens the time-pressure curve, which in turn changes the "timing."

With the exception of an increase in muzzle velocity, it does not seem to have any advantage except, possibly, for extremely cold temperatures, very slow-burning powders or very short barrels with powders burning too slow for the length of the barrel. Do not overlook the physical danger, either. To be competent, the reloader would have to start off his experiments with one or two grains of black powder or Pyrodex, while using a constant charge of a smokeless powder, and incrementally work up his load at 1.0 grain intervals until he finds the correct combination, a combination that must correspond with the correct muzzle velocity, timing, and rate of twist.

While, on the other hand, a few grains of black powder or Pyrodex (with others in the market) between the primer and a relatively slow-burning powder may not only increase the ignition temperature and shorten the time-pressure curve but make the powder behave as if it were a much faster powder. It allows for a higher muzzle velocity, in a small remaining case capacity, with very slow pistol or very fast-burning rifle powders that are ordinarily not really suitable for this application.

2.8 SOME SUGGESTIONS

Going back to Groups VII to X of Table 2.3, it is obvious, if we want to drop the size of the groups from the size of a man's face to – say, a man's nose – we will need to do some work to improve the ballistic uniformity of the ammunition. *Ballistic efficiency* is of course extremely important.

Ballistic uniformity is more important, moreover. Without both efficiency and uniformity, however, we cannot seriously expect to consistently hit our targets. Grouping is everything!

To help the reloader, we offer the following advice when using a revolver. Needless to say, most of this advice would apply to any kind of firearm, not just handguns or revolvers. We are using a revolver to exemplify the kind of work and experimentation we would need to carry out in order to "tune up" a gun. With some important exceptions, these exemplications and experimentations apply to almost any gun. If it does not have a cylinder or is a single-shot affair, then we can discount some of the following advice, obviously. On the most part, the following advice would apply to nearly every gun.

- Perhaps the revolver's cylinder may need smoother and rounder chambers with a closer tolerance in respect to each other? If one chamber is one or two thousands of an inch off the other chambers, it is imperative to know that and then to isolate it from the other cylinders. Some serious shooters shooting with revolvers in competition have been known to take their cylinders to a jeweler to engrave a number for each chamber on the extractor in order to keep track of the individual performance characteristics of each chamber. Others simply go to the manufacturer to purchase a better cylinder.
- Check the cylinder's center for trueness. If it is off center, each chamber will possess its own personality in respect to the barrel and forcing cone. Each bullet from each chamber will not enter into the forcing force uniformly. Ballistic *uniformity* becomes impossible. The only solution is to get a new and truer cylinder.
- Make sure the center of each chamber is true to the center of the cylinder. If not, the only solution is to get a new cylinder.
- If the barrel accumulates lead in the bore at the correct muzzle velocity for the rate of twist and with the proper lubricant, there are probably burrs in the rifling. Lap the bore. Lapping compound is readily available in the market.
- Measure the outside diameter of each bullet with a micrometer. Put them into lots representing diameter. Use the same precise diameter

in five or six shot lots for grouping purposes. This will provide superior results downrange, because every bullet will have precisely the same outside diameter. Do the same thing for every caliber and gun.

- Make sure each bullet weighs the same and do the same thing, as you did with diameter, by placing the same weights into lots for uniformity.
- After resizing the lead or jacketed bullets, if the bullets vary in diameter by more than one to three thousands of an inch, as we rotate the micrometer around the bullet for at least three separate readings, we must either get a new die or do the same thing as above. Place them into lots. This will greatly improve accuracy downrange. Remember! Uniformity!
- Check the base of each bullet. If unable to get perfect bases, with the same diameter and roundness, if casting lead bullets, then get a nose-fed bullet mold. If the bullet bases are not perfectly round, they will not shoot accurately, either. The roundness of the bullet's base is probably the most important attribute of a bullet and will affect accuracy more than any other bullet attribute.
- Check the cartridge cases. If they are not of the same brand and lot number, they will not shoot uniformly if we were to mix them together in a group. Every manufacturing brand and every lot of every brand will have its own personality. Make sure, when shooting in groups for accuracy, we do not mix together different brands and lots.
- Measure the length of each cartridge case. If any case in the same lot of any manufacturing brand varies by as much as one thousands of an inch, it will affect the nature or strength of the crimp and hence the "timing" of the bullet traveling through the bore. Make its length as close to each other as physically possible. As long as each case, in a given lot, is the same as each other, the actual length is unimportant. Strive for uniformity.
- Standardize reloading techniques and methods and stick to them. A variation, however small, in reloading techniques will affect the "timing" if it should occur while in the middle of a group or lot.

2.9 CONCLUSION

In this chapter, although we could have easily started off with other guns and calibers, we decided to use a .44 Magnum Smith & Wesson Model 29 revolver instead as a representation of guns and the various ballistic problems and methods of resolving them. With the right load, this gun and caliber will shoot very accurately within its inherent design limitations, the bullets' maximum effective range and maximum range of lethality for its weight and trajectory velocity.

When we began to examine the gun ballistically, we discovered the rate of twist dictated a very slow muzzle velocity of 835 fps with the typical 240 grain bullet in sharp contrast to the ammunition commercially available on the market. Depending on the barrel's length, those bullets will usually leave the barrel in excess of 1,000 fps with flat soft nose-jacketed bullets. Even with a cast lead-alloy bullet, the loads are designed for close to 1,000 fps using gas checks on the base of the bullet to allow for those velocities without excessively "leading" the bore.

The general public wants it that way and manufacturers will give them what they want even if wrong.

As demonstrated in this text, those muzzle velocities with commercial ammunition are simply too fast to be ballistically compatible with the gun's rate of twist. As stated before, regardless of the gun, rifle, or handgun, with the exception of military weapons, the rates of twist are almost always wrong with most bullets.

Military firearms are designed to use only one bullet weight and the rate of twist selected for that weight; commercial firearms, on the other hand, although designed in the same way, are used by the shooting community with different weights and lengths for the same cartridge.

It is usually necessary, if we want to obtain the optimum accuracy for a particular gun, to reload our own ammunition, once we have calculated the correct muzzle velocity for a given rate of twist and bullet length.

We decided to stay with the 240 grain cast lead-alloy bullet and use a relatively fast-burning shotgun powder in an experiment to identify the amount of Red Dot it would take to produce the correct muzzle velocity of 835 fps or close to it. We knew it would be almost impossible or at least very difficult to consistently obtain that exact muzzle velocity, shot after shot. There are simply too many variables to control; that we demonstrated in this text with our experiments.

As far as the decision to use Red Dot, with more than 100 gunpowders available on the market, that decision required extensive experience with both reloading ammunition and shooting guns, the kind of experience that takes years to accumulate. It is an intuitive process – not analytical – and also an intellectual process our brains develop over time to avoid another slow exploratory procedure we needed to do originally, when we were more naïve and less knowledgeable, in order to derive one of many correct solutions in much less time and mental labor. It only has to be a correct working solution, out of the many possible correct working solutions. It does not have to be the only correct working solution, either, nor excluding possibly one of the better working solutions. It is just a solution that works.

To improve the ballistic efficiency, we used corn cob as the filler to fill up the remaining case capacity on top of the Red Dot. Corn cob increased the muzzle velocity by approximately 50 fps, stretched out the time-pressure, slowed the burning rate characteristics of the powder, and changed the "timing" to more closely correspond with the length of the barrel.

In the real world, however, we discovered muzzle velocity changes constantly with the constant change in air temperature. When the temperature was a little over 90°F during a day of July 1986, the average muzzle velocity was 841 fps with 5.2 grains of Red Dot and 8.5 grains of corn cob. In the following September, with a much lower air temperature, the same combination produced a drop in muzzle velocity by an average of 88 fps.

When we re-tested the same load later in November of that year, with the air temperature of 34°F, muzzle velocity dropped a total of 151 fps from the July reading of 841 fps to a new average of 690 fps.

Clearly, we could not rely upon one particular load to maintain a constant muzzle velocity throughout a year with seasonable temperature changes, too. Group XI demonstrates the extreme variability in muzzle velocity with temperature changes.

In Group XI of Table 2.4, with an outside air temperature of 34°F, shooting a load of 5.2 grains of Red Dot without the filler, we obtained an average muzzle velocity of 690 fps.

TABLE 2.4

Group XI	Group XII
5.2	5.2
No filler	Corn Cob filler
1. 677 Lo=677	1. 816 Lo=728
2. 682 Hi=714	2. 728 Hi=816
3. 701 Av=690	3. 730 Av=754
4. 714 ES=37	4. 740 ES=88
5. 680 Sd=16	5. 756 Sd=36
Group XIII	**Group XIV**
5.2	5.4
Corn Meal filler	No filler
1. 785 Lo=772	1. 727 Lo=684
2. 772 Hi=836	2. 716 Hi=733
3. 836 Av=797	3. 733 Av=717
4. 802 ES=64	4. 726 ES=49
5. 794 Sd=24	5. 684 Sd=19
Group XV	**Group XVI**
5.4	5.4
Corn Cob filler	Corn Meal filler
1. 792 Lo=769	1. 818 Lo=812
2. 806 Hi=806	2. 817 Hi=819
3. 784 Av=789	3. 819 Av=816
4. 769 ES=37	4. 814 ES=7
5. 796 Sd=13	5. 812 Sd=3

With the same load, but using corn cob filler, we increased muzzle velocity by 64 fps to an average of 754 fps (see Group XII of Table 2.4).

Then, when we changed the filler to corn meal in Group XIII, we increased muzzle velocity by another 43 fps to an average of 797 fps.

These two fillers clearly work to increase muzzle velocity in these experiments.

Please take the time to observe Group XII more closely. Look at the muzzle velocity of 754 fps, which had been obtained with an outside temperature of 34°F. Now compare this fact to the average muzzle velocity obtained in Groups VII to X of Table 2.3, the previous September with an average outside temperature of 30–35°F – much higher. Without the corn cob filler, the average muzzle velocities varied with the temperature from July to November, an extreme variation from 90–94°F to a cold 34°F. Yet, the same loads with corn cob filler maintained a *constant* medium muzzle velocity of (actually) 753.5 fps.

What is the logical conclusion? The corn cob filler – possibly other fillers as well – can maintain a constant muzzle velocity while temperature changes from one extreme to another.

Notice, as a final observation in this chapter, we can find some interesting performance characteristics in Group XVI. With a load of 5.4 grains of Red Dot and corn meal filler, the extreme spread was 7 fps, an SD of 3 fps (actually 2.6 fps) and an average muzzle velocity of 816 fps.

These experiments clearly demonstrate, though definitely not definitive by any stretch of an imagination, that corn meal not only increases muzzle velocity on an average of 100 fps in this cartridge and caliber, but actually reduces the extreme spread (see Group X in Table 2.3), improves the ballistic efficiency and uniformity, and, all the while, maintains a more constant muzzle velocity and with a variable temperature in the real world.

CHAPTER 3

A THEORY OF THE ASYMPTOTIC FUNCTION

CONTENTS

"*The word asymptote is derived from the* ***Greek*** ἀσύμπτωτος (*asumptōtos*) *which means 'not falling together.' The term was introduced by* ***Apollonius of Perga*** *in his work on* ***conic sections****, but in contrast to its modern meaning, he used it to mean any line that does not intersect the given curve.*"

—Wikipedia, the free encyclopedia

3.1 INTRODUCTION

A long, long time ago, when mathematicians began to play with numbers, they recognized even then the development of patterns and relationships seemingly corresponding with their daily experiences of life and human reality.

In some ages of the ancient past, mathematicians – perhaps as a corollary of their cultural environment – assigned magical or supernatural values to numbers and, on occasion, killed and even murdered people to protect themselves from the people outside trying to discover "truth and knowledge" through numbers. Indeed, we even have documented accounts of people committing suicide to protect them from revealing these magical truths of numbers to people outside of their congregation.

So, over these long and eventful centuries, we have an equally long history of interesting stories of either the establishment or destruction of religious, pseudo-religious, and philosophical groups of people organized within the framework of the magical or supernatural interpretations of the patterns and relationships of numbers.

Even today, with our enormous knowledge databases and technologies, we continue to assign magical values and special significances to numbers and perhaps, too, within the framework of the folkways and mores of our own cultural environments.

Without our getting into a religion of mathematics, some new or old philosophy or perhaps a line of continuity on nonsense dating back to prehistory, the serious and mature student of the science of ballistics must eventually recognize the existence of patterns and relationships between the arrangements of numbers to ballistic phenomena.

Let us take the theory of the asymptotic function as an example.

About approximately 39 years ago at Camp Curtis Guild in 1976, near Boston, Massachusetts, we began to perceive an apparent relationship between the arrangement of certain numbers to the temperature of a gun barrel and its effect toward the accuracy of a bullet striking a target downrange. We noticed the same relationships while performing experiments at the Westfield Sportsmen Club in Westfield, Massachusetts, between 1978 and 1980.

As the barrel warmed from each successive shot, the groups downrange became successively smaller until they had obtained their maximum potential for accuracy for that particular gun and ammunition. At the same time, the groups would move with the increase in barrel temperature from the bottom right to the center of the target.

Through our intuitive powers, we thought we could perceive a relationship between the severities of this climb to the center of the target to the temperature of the barrel.

When the temperature of the barrel was cold and well below 50°F, the first shot was always low and, in our location above the *equator*, to the right of center (see coriolis effect).

After each successive shot, as the barrel warmed and approached 70°F, not only did the groups grow smaller as they climbed to the center of the target, each successive shot produced a successively smaller increment of climb. Each group and each increment of climb grew smaller and smaller as the barrel became hotter and hotter.

Additionally, once the tendency to climb subsided and settled down to the center of the target, as the barrel grew even hotter and exceeded 100°F, the group placement shifted around the center until the barrel had reached an even higher temperature. Then, at that point, the groups settled down again.

Later, when the barrel became so hot, when ordinary gun lubricants and solvents would vaporize to the touch, we could observe the same phenomena of groups shifting relative to the target's center before settling down again as the barrel reached a new high temperature.

As each of us ought to know very well, or by now, the use of our intuitive powers can be dangerous, particularly when we must consider a long history of people deriving conclusions and making decisions intuitively within the context of their psychiatric issues and cultural biases. We can look at Adolph Hitler, as an example of a man, who thought of himself as a genius and was diagnosed as a paranoid psychotic. Some historians describe him as an "evil genius."

Nevertheless, his psychosis consistently distorted his perception of reality to such a degree that the Ally commanders during WW II made a conscious decision not to assassinate him out of fear that, if he were killed, a more competent replacement commander would fight a better war against them.

Nonetheless, intuition can be extremely useful for us if we were to use it conscientiously. It can allow us to jump over enormous chunks of data, sometimes even instantaneously, to obtain conclusions and insights when, ordinarily, systems of analysis and logic would require years of methodical research to perform correctly.

3.2 THEORY

The asymptotic function in mathematics refers to a line extending from a maximum value on a sheet of graph paper and moving to a successively smaller value at each subsequent increment. However small each final value may represent, it will never obtain a value of zero.

If we were to take any number, say "1," and to divide it by another number, say "x," we would get a number smaller than "1." Then, if we were to allow "x" to grow progressively at each increment of computation, we would get a characteristic curve similar to the above curve in Figure 3.1. We call this mathematical phenomenon the "*Asymptotic Function.*"

$$Z = 1/x \tag{1}$$

Let's work out an example in exhaustive detail. Convert the number "1" to the number "200." Let "x" represent temperature at increments of 10°F from 0°F to 510°F.

- 200 divided by "0" is "0."
- 200 divided by "10" is "20."
- 200 divided by "20" is "10," etc.

When we finish our computations, we should get the results in the *second column* of Table 3.1.

However, though we can barely perceive a pattern in the areas identified as Groups A, B, C, etc., we really need a further amplification to

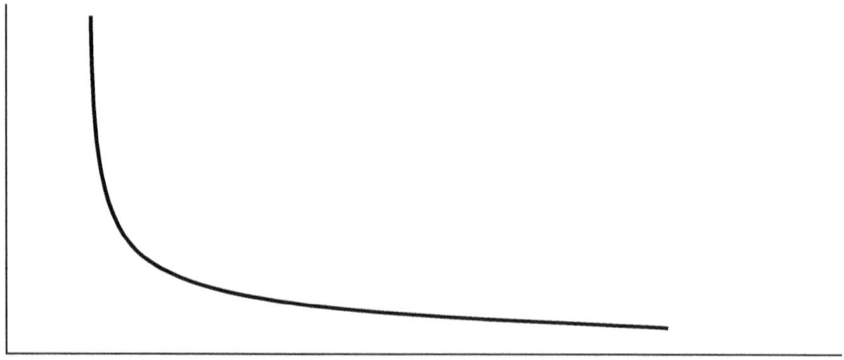

FIGURE 3.1 The asymptotic function characteristic curve.

TABLE 3.1 An Asymptotic Representation of Temperature

Temp. °F	200/T°	U-L	A		Temp. °F	200/T°	200/T°	U-L	A
0°	0	0	0		260°	.77	.03	6	
10°	20	20	4000		270°	.74	.03	6	Group F
20°	10	10	2000		280°	.71	.02	6	
30°	6.7	3.3	663		290°	.69	.02	4	
40°	5	1.7	340		300°	.67	.02	4	
50°	4	1.0	200		310°	.65	.02	4	
60°	3.3	.07	139		320°	.63	.02	4	
70°	2.9	.4	81		330°	.61	.02	4	
80°	2.5	.4	80	Group A	340°	.59	.02	4	Group C
90°	2.2	.4	79		350°	.57	.02	4	
100°	2	.2	40	Group B	360°	.56	.01	4	
110°	1.8	.2	40		370°	.54	.02	4	
120°	1.7	.1	20		380°	.53	.01	4	
130°	1.5	.2	20		390°	.51	.02	4	
140°	1.4	.1	20		400°	.50	.01	2	
150°	1.3	.1	20	Group C	410°	.49	.01	2	
160°	1.25	.05	20		420°	.48	.01	2	
170°	1.2	.1	20		430°	.47	.01	2	
180°	1.1	.1	20		440°	.46	.02	2	
190°	1.05	.05	10	Group D	450°	.44	.01	2	Group H
200°	1.0	.05	10		460°	.43	.01	2	
210°	0.95	.05	8		470°	.43	.009	2	
220°	0.91	.04	8	Group E	480°	.42	.01	2	
230°	0.87	.04	8		490°	.41	.01	2	
240°	0.83	.04	8		500°	.40	.01	2	
250°	0.80	.03	6		510°	.40	.008	2	

identify the real beginning and end of each pattern. We can obtain this amplification with a simple manipulation. When we multiply the asymptote in the *second column* by the difference between the preceding asymptote (U) and the present asymptote (L), we obtain a clarification of these patterns in the *third column*.

Still, there is confusion between the beginning and the end of each pattern. So, again, we must perform yet another simple manipulation to sharpen the distinction between the two. When we multiply the answer in the *third column* by the temperature in the *first column*, we finally obtain something meaningful in the *fourth column*.

Now, we can easily perceive and recognize each pattern identified in Table 3.1 as Groups A through H.

Thus, the asymptotic representation of temperature (Æ) takes the following mathematical expression:

$$Æ = (PN/T°)(U\text{-}L)T° \tag{2}$$

where Æ = the asymptotic representation of temperature, *PN* = the "priming number," and, in this instance, represent 200 (see Table 3.1).

Actually, *PN* can represent any number, from 1 to infinity; however, as we increase or decrease *PN*, we increase or decrease the distinction between the patterns, as well as the size of the numbers and the number of patterns. We must also "round off" each number to the nearest whole number or we will have destroyed their distinctions and thus complicate the interpretation of the patterns.

Please refer to the first column of Table 3.1. Suppose, for the moment, we were to fire a gun using a barrel with a temperature of 10°F. As we can see in the fourth column, the asymptotic representation of 10°F is 4000 units.

Of course, once we will have fired the first shot, the barrel immediately warms up to a higher temperature. For the sake of convenience in this example, suppose the new barrel temperature becomes 20°F, which gives us an asymptotic representation of 2000 units, an enormous drop by exactly 50%. If this asymptotic representation corresponds with ballistic reality, then we can expect a significant and radical shift in the placement of the second shot relative to the placement of the first shot.

Likewise, if the third shot were to warm the barrel to 30°F, the asymptotic representation drops to 663 units – but, this time, much more than 50%. Once more, we would expect a significant shift in the placement of the shot relative to the two preceding shots, and so forth.

After that, at 40°F, we notice a reduction in the drop of the asymptotic representation. Instead of a successive increase in the drop of asymptotic units as temperature increases, there is now a tendency to slow down its rate of decrease. Look at the characteristic curve in Figure 3.2.

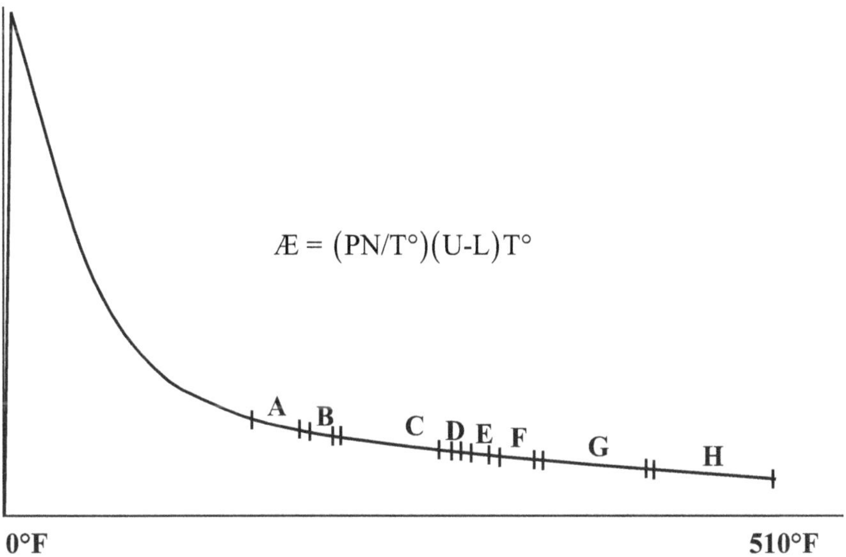

FIGURE 3.2 An asymptotic representation of temperature – a characteristic curve.

As soon as the barrel warms to 70°F, the asymptotic representation stabilizes for a temperature range of 30°F or until it reaches 90°F. It follows that, at 100°F, there is another significant drop in the asymptotic representation by 50%. This fact leads us to the inescapable conclusion that, if a barrel were warming from 90°F to 100°F, there should be a highly noticeable shift and disruption in the placement of shots downrange and in the size of the groups as well.

Stabilization would not reoccur until the barrel reaches a temperature of 120°F; then another shift in the placement of shots and grouping should become evident at 190°F.

3.3 APPLICATION TO REALITY

The greatest stabilization of both placement and grouping would seemingly occur in Group C, a temperature range from 120°F to 180°F, or a total of 60°F in bandwidth. This suggests that, if we could design a gun barrel to maintain a constant medium temperature of 150°F, then a fluctuation in temperature from shot-to-shot, if we could keep it between 120°F and 180°F, would reflect no noticeable change in either the placement of shots or the size of groups downrange. The first shot would have the same placement downrange as the last shot, with no shifting or climbing tendencies. Groups would stay constant.

Better still, if we could maintain a *minimum* barrel temperature of 120°F, even if each successive shot were to increase the barrel temperature, as long as the *maximum* temperature never exceeds 180°F, both placement and grouping should be constant downrange.

Figure 3.1 should help us to conceptualize this asymptotic representation of temperature and its immediate engineering application. Stabilization of shot-placement and grouping corresponds with relatively flat and straight positions on the characteristic curve. Nor until the barrel reaches a temperature of nearly 70°F does the curve nearly flatten out. When the barrel is colder, between 0°F and 60°F, each increment is massive and disruptive as the temperature increases with each successive shot. If we can keep the barrel's temperature between 120°F and 180°F, however, the relatively flat curve at this bandwidth will provide a smooth transition with the increase in temperature after each shot.

3.4 CONCLUSION

If our intuition proves correct, then the asymptotic representation of temperature becomes an important consideration toward the study of ballistics and the mechanics of accuracy.

If we were to design a gun barrel to maintain a constant initial shooting temperature within an interpretation of the asymptotic representation of temperature, we could very well guarantee the probability of a first-shot hit downrange and then maintain, perhaps indefinitely, all subsequent placements and groups irrespective of the time between shots.

Asymptotic function is no longer a mathematical abstraction. It has a relationship to the science of ballistic phenomena and at least one engineering application concerning the study, design, and development of long-range guns and other ballistic weapons as well.

Then, again, if we were to think about it for a moment, this asymptotic representation of temperature may explain other physical phenomena as well, including the load-bearing characteristics of an automobile battery in the cold of winter.

Apollonius of Perga (c. 262 BC–c. 190 BC) was a Greek geometer and astronomer who wrote on conic section to greatly influence Ptolemy, Francesco Maurolico, Johannes Kepler, Isaac Newton, and René Descartes. He gave us the ellipse, the parabola and the hyperbola, among many other developments, such as asymptote, including an explanation of the motion of the planets and the different speeds of the Moon.

CHAPTER 4

THE THEORY OF TWIST

CONTENTS

If we were to increase muzzle velocity, for a given bullet length and diameter, it would become necessary to *decrease* the rate of twist in the barrel in order to properly stabilize the bullet.

If we were to increase the length of the bullet, from its base to its tip, without either increasing its diameter or muzzle velocity, we would need to *increase* the rate of twist to stabilize the bullet.

Therefore, there is now enough information to describe this relationship between bullet lengths, bullet diameter, and muzzle velocity in a mathematical formula.

$$TW = 2d\sqrt{[MV/(Bl/d)]} = 2\sqrt{d^3(MV/Bl)} \qquad (1)$$

where TW = the proper rate of twist (turns per inch); d = the bullet's diameter (inch); MV = muzzle velocity (feet per second); Bl = bullet length (inch).

On the other hand, if we were to place a boat-tail configuration at the base of the bullet, as is common among American military ballistic small arms and artillery projectiles, it would have the effect of actually *decreasing* the length of the bullet and *decreasing* cylinder drag; then we would need to multiply the rate of twist by a factor of 0.90 to *increase* the rate of twist.

$$TW = TW \times 0.90 \tag{2}$$

From the above relationships and formulas, we can derive the *best muzzle velocity* (BMV), for a given rate of twist, bullet diameter, and length, with the following algebraic manipulation.

$$\text{BMV} = \frac{\frac{1}{2}\left[(Bl/d)TW^2\right]}{2(d)2} = \frac{Bl\left(TW^2\right)}{4d^3} \tag{3}$$

From that equation, we can just as easily calculate the *best bullet length* (BBL) with a given rate of twist, muzzle velocity, and diameter, as:

$$\text{BBL} = \frac{2\ (MV \times d)}{TW^2}\ 2(\text{d})^2 = \frac{4\ (MV \times d^3)}{TW^2} \tag{4}$$

4.1 THE LAWS OF TWIST

If we were to increase the bullet's weight/length, without a change in muzzle velocity, we would need to *increase* the rate of twist.

If, on the other hand, we were to increase muzzle velocity, without an increase in the bullet/ length, we would need to *decrease* the rate of twist.

Then, if we were to increase both the bullet's muzzle velocity and weight/length, we would need to *increase* its rate of twist.

For a given rate of twist, when we increase muzzle velocity, we would need to *increase* the bullet's weight/length to properly stabilize it.

For a given rate of twist, if we were to decrease muzzle velocity, we would need to *decrease* the bullet's weight/length to properly stabilize it.

CHAPTER 5

THE THEORY OF BULLET SPIN

CONTENTS

5.1 INTRODUCTION

Some time near the end of the 15th century, some people began to recognize the importance of causing the bullet to spin in order to properly stabilize it. Improvements were made by August Kotter in 1520 (biographical material unavailable), who was an armorer in Nuremberg, Germany, at the time; the original person responsible for the idea is unknown or uncertain. Though several methods are presently available to cause the bullet to spin, some of them more effective than others, with the most effective method – for small arm ballistics – has been to install "rifling" inside the bore of the barrel. As the bullet moves through the bore, the bore pressure forces the bullet to take the shape of the rifling to spin at the rate of the rifling twist in the bore.

Spin is directly proportional to velocity (V) and the rate of twist (TW). If we increase the bore velocity, we increase the rate of spin. If we increase the rate of twist, then we increase the rate of spin ($\hat{w}$).

Therefore, we have the following mathematical relationship between velocity and twist to spin.

$$\hat{w} = V \times TW \quad (1)$$

Under either an experimental or field condition, if we know the bullet's velocity and rate of spin, we can determine the barrel's rate of twist with the following equation:

$$TW = \frac{\hat{w}}{V} \qquad (2)$$

Then, if for some reason, if we need to know velocity but know the bullet's rate of spin and the barrel's rate of twist instead, we can calculate its velocity with the following equation:

$$V = \frac{\hat{w}}{TW} \qquad (3)$$

A study of these relationships will reveal some interesting phenomena:
– If we can properly stabilize the bullet in the bore, with the proper spin, the bullet will always have the proper stability at any distance downrange. As the bullet's velocity decelerates in its flight path, the rate of spin also decelerates proportionately, which tends to keep the bullet at the proper rate of spin for the proper stabilization (see Eq. (7)).

5.2 CONCLUSION

When we increase muzzle velocity, we increase the rate of spin. When we increase the rate of twist, we also increase the rate of spin. If we can properly stabilize the bullet with the proper rate of spin, the bullet will always maintain the correct rate of spin to correspond with the deceleration of velocity throughout its entire flight path. So, it is critically important to use the proper rate of twist. It makes a big difference.

CHAPTER 6

THE THEORY OF KINETIC ENERGY

CONTENTS

"According to physics, kinetic energy is one of many types of energy that exist. This is energy generated because something is moving – the faster it's going, the more kinetic energy it has. A person sitting has no kinetic energy, but a person running like a maniac has tremendous kinetic energy: if they run into you, you'll feel the brunt of it. Footballs, baseballs, rocks, bullets, airplanes, and anything else moving quickly through the air all have kinetic energy."

—Vocabulary.com

Albert Einstein holds the credit for being the first person responsible for working out the mathematical relationship between mass and velocity to energy in subatomic particles with his famous equation of $E = MC^2$, where E = energy; M = mass; and $E = MC^2$ means the kinetic energy of moving mass is mass × mass × the square of its velocity. C^2 means the square of the mass's velocity at the speed of light. This equation deals with the relationship between subatomic particles, such as the mass of light travelling at the speed of light. It does not address the problems of bullets or projectiles traveling at velocities substantially below the speed of light.

"Albert Einstein (14 March 1879–18 April 1955) was a German-born theoretical physicist. Einstein's work is also known for its influence on the philosophy of science. He developed the general theory of relativity, one of the two pillars of modern physics (alongside quantum mechanics). Einstein is best known in popular culture for his mass–energy equivalence formula $E = mc^2$ (which has been dubbed "the world's most famous equation"). He received the 1921 Nobel Prize in Physics for his "services to theoretical physics," in particular his discovery of the law of the photoelectric effect, a pivotal step in the evolution of quantum theory."

—Wikipedia, the free encyclopedia

It was in the early 19th century, with people such as *Gaspard-Gustave de Coriolis or Gustave Coriolis* (May 21, 1792–September 19, 1843), a French mathematician, mechanical engineer, and scientist, who may have been the first person who presented mechanics in a way people could understand and use, and that lead to his development of the notion of kinetic energy.

He published a book entitled, *Calcul de l'Effet des Machines* ("Calculation of the Effect of Machines") and, during this period, the theory of kinetic energy of something moving well below the speed of light, such as bullets and artillery projectiles, became recognized as $^1/_2 MV^2$.

William Thomson, (June 26, 1824–December 17, 1907), later Lord Kelvin, has the credit for inventing the term "*kinetic energy*." Kinetic energy is the consequence of mass in motion. Without motion, there is no kinetic energy, only *potential energy*.

Over the years, some people had recognized these two equations as inadequate when dealing with anything moving less than but substantially near the speed of light. So, over time, modifications of that brilliant formula by Einstein and earlier works by Coriolis and Lord Kelvin were made to accommodate the realities of the growing field of physics. We now have several modifications to calculate the kinetic energy of particles travelling at, near, below, or substantially below the speed of light, including ballistic projectiles travelling only a few thousand feet a second.

Because small arm ballistics is the study of not particles or the mass of light but of relatively large projectiles travelling at very slow velocities, relative to the speed of light, we must modify the equation further to accommodate the realities of ballistics. We use the following equation, worked out originally in the early 19th century for the science of mechanics in England's 19th century industries to calculate kinetic energy of bullets in small arm ballistics:

$$KE = \tfrac{1}{2} MV^2 \qquad (1)$$

As we can easily see, the above equation is a slight modification of Einstein's $E = MC^2$, but has not changed since the early 19th century, in which we multiply the bullet's mass by the square of its velocity and then divide the answer by 2.

In small arm ballistics in the United States, however, we modify the equation further:

$$KE = \frac{(W / 225218)V^2}{2} \qquad (2)$$

where W = Weight (grains); V = Velocity (fps)

The figure, 225218, represents the product of one Avoirdupois pound (7000 grains) and acceleration due to gravity or 32.172 ft/second/second, designed to convert the weight of the bullet to its mass. In the English unit of measurement, we calculate mass by dividing the weight of the bullet, in grains, by the product of 7000 and 32.174 (which equals 225218).

The figure, 32.172, is the *standard* in the American firearms industry for acceleration due to gravity in the calculation of trajectory for the ballistic tables common today among members of the shooting community, and among bullet manufacturers in their reloading manuals and ballistic tables, a figure representing the middle-ground in the above Table 6.1.

Obviously, Table 6.1 clearly reveals, as we move from either the North or the South Pole to the Equator, acceleration due to gravity increases from a minimum of 32.0862 feet per second from either pole to 32.2575 feet per second at the equator. As we approach the equator, from either pole, things falling from the sky fall faster (accelerates) – including small bullets from guns.

A bullet of a given weight falling from the sky over the North Pole would possess less velocity and kinetic energy than the same bullet falling from the sky directly over the equator, though the difference is extremely small in its total value but important in the accurate calculation of a bullet's flight path or trajectory in any given position on the planet Earth.

If we were to add up each value of acceleration due to gravity in Table 7 and divide by 10, we would get the average value of 32.17187 or 32.172; this is the reason behind the standard use of 32.172 in the firearms industry and shooting community.

6.1 THE EFFECT OF BULLET SPIN OVER ANGULAR KINETIC ENERGY

Many people in the shooting community, thinking independently of each other, have developed the same conclusion – the standard equation to

TABLE 6.1

Acceleration due to Gravity (g)
0° – 32.0862
10° – 32.0916
20° – 32.1062
30° – 32.1290
40° – 32.1570
50° – 32.1867
60° – 32.2147
70° – 32.2375
80° – 32.2523
90° – 32.2575

calculate kinetic energy is defective! It fails to consider the kinetic energy of the spin the bullet invariably creates; however, some of them – and it happens almost every day – after careful consideration and research at their local library or on the Internet in Wikipedia, the free encyclopedia, they find the bullet's spin generates a negligible amount of kinetic energy, either in its forward direction or in the amount it transfers into each increment of penetration of its flight path or in the target. It is truly a trifling amount, though a value we can easily calculate if important.
In any event, for those who want to know, the relationship between linear (forward) and angular (spin) kinetic energy is the sum of linear kinetic energy and one-half of the product of the bullet's moment of inertia (I) and the square of the angular velocity ($\hat{w}$).

$$KE = \tfrac{1}{2}MV + \tfrac{1}{2}I\hat{w}^2 \tag{3}$$

For round balls in the muzzle loading community, the same equation takes the following appearance:

$$KE = \frac{\left[MV^2 + \left(2Mr^2/5\right)\right]\left[V\left(\pi d/TW\right)\right]^2}{2} \tag{4}$$

where "r" is the bullet's radius; "M" represents mass of the bullet; "π" represents "Pi" or 3.1415 and "d" is the bullet's diameter. The above equation only works with the English unit of measurement. For the metric system, we would have to modify it, of course.

6.2 CONCLUSION

Kinetic energy is the relationship between the bullet's mass and its velocity. The greater the mass, for a given velocity, the greater is the kinetic energy; and the greater the bullet's velocity, for a given mass, the greater is the kinetic energy in the bullet as it transfer a small increment of its kinetic energy into each increment of penetration of its flight path.

The American shooting community uses the English unit of measurement in its calculations of bullets travelling through 3,600 cubic inches in its flight path for every interval of 100 yards following a parabolic trajectory due to acceleration due to gravity. It cannot travel in a straight line because of gravity, hence the parabolic trajectory.

On its way through this flight path, moreover, it transfers a small amount of its kinetic energy into each cubic inch of penetration, a product of transfer of energy (*TE*) and air density (σ), or $TE = KE \times \sigma$ (in this text, we will stay with dry air density of 0.0000509 lb/cu./in. for our calculations). To be much more accurate, of course, we must use damp air density.

When there is a transfer of kinetic energy from the bullet into the air of its flight path, it subtracts that amount of transfer of kinetic energy from the kinetic energy of the bullet travelling forward to produce a slight reduction in its forward velocity. With a reduction of that forward velocity, without a loss in bullet mass, there is a corresponding slight reduction in kinetic energy, where $KE = \frac{1}{2} MV^2$. This reduction or loss of forward kinetic energy adds up and represents one of the three variables responsible for the bullet's drag: drag induced by transfer of energy; drag induced by the surface area of the bullet's nose; and drag induced by the surface area of the bullet's cylinder.

When the bullet arrives at the target, it has an initial terminal velocity (ξ) that represents the lethal kinetic energy going into penetration (or reflection), and that amount is the amount remaining after the accumulative transfer of energy into each increment of penetration in its flight path. That accumulative amount is the summation of the total amount transferred into each cubic inch of penetration of the flight path before entering the target.

Unlike light, which travels through a path with density, pursuant to the transfer of energy relationship, transfers an increment of its forward kinetic energy into each increment of penetration without losing forward velocity or kinetic energy, a bullet in free flight, however, loses both velocity and kinetic energy at each increment of penetration and, eventually, even if it does not strike anything, loses all forward velocity to fall to the ground due to gravity.

Light, on the other hand, does not lose any of its forward velocity from transfer of energy unless the density of the path it travels changes; then, it will increase its forward velocity and energy with a reduction of density in its path and decrease its forward velocity and energy with an increase in the density – not from transfer of energy.

It will go up and down with its forward velocity and energy, and transfer energy into each increment in its path, as the path's density goes up and down. Interestingly, unlike a bullet, an electrostatic charge in the light's path modulates the light's velocity to make it also go up and down with the strength of its charge. Bullets in free flight will not do that.

A bullet in free flight loses it velocity and kinetic energy due to the drag of transfer of energy, drag induced by the surface area of the nose and drag induced by the surface area of the cylinder.

Light, in sharp contrast, loses energy in a relationship of one/half of the cube root of the ratio of the energy to the third power of the distance between the source of the energy and the target downrange or:

$$Em = \tfrac{1}{2}\sqrt[3]{\mathrm{E} / \mathrm{D}^3} \tag{5}$$

where Em = remaining energy at the point of measurement (joules/sq.cm/second); E = energy output from the source (Joules/second); D = distance between the source of the energy and the target downrange (kilometers).

We can use a common flashlight as an analogy to explain the way light loses its energy through dispersion. Go into a darken room and, at an approximate distance of two feet, turn on the flashlight facing a wall. Then slowly back up until the diameter of the projected illumination on the wall gets so large it disappears altogether. Energy from our Sun does the same thing but at a much greater distance due to its enormous power output of approximately 3.86×10^{33} ergs per second or 3.86×10^{28} joules per second, but on the outer surface of the Earth's atmosphere, about 93 million miles away, the energy level drops to a mere 8.331732 joules per second per square centimeter.

Acceleration due to gravity and the velocity of things and bullets falling from the sky vary with the latitudinal position on our planet. As we move from either the North or South Pole to the equator, acceleration due to gravity increases, from the lowest value of 32.0862 at either the North or South Pole, to a maximum value of 32.2575 at the equator. So, the weight of things and bullets vary with its position relative to our planet's latitudinal lines. Velocity and kinetic energy increases or decreases as we move in between the poles and the equator.

Angular (spin) kinetic energy is too insignificant for us to include in the calculation of kinetic energy, ballistic tables, or the calculation and prediction of trajectory in small arm ballistics. So, we ignore it.

CHAPTER 7

TEMPERATURE CONVERSION FORMULAS

CONTENTS

7.1 INTRODUCTION

Frequently, it has become necessary for us in the shooting community to convert a temperature reading from one scale to another.

The International System of Units is a name adopted by the Eleventh General Conference on Weights and Measures, which was held in Paris, France, in 1960, in order to develop a consistent system of units of measurements based on the meter-kilogram-second system (MKS).

This international system is called SI and represents the initials of System International.

In the same year, the Conference adopted a temperature scale based on a fixed temperature, namely the triple point of water in which the solid, liquid, and gas maintain an equilibrium.

In the Fahrenheit scale, the system is based on 32°F as representing the freezing point of water and 212°F as its boiling point. While, on the other hand, the Celsius scale uses 0°C as the freezing point and 100°C as the boiling point.

However, we can use a mathematical formula to convert one to the other, or back. Actually, we have eight conversion formulas.

(1) °C to °F = (°C × 9/5) + 32
(2) °F to °C = (°F – 32) × 5/9
(3) °C to °K = °C + 273.15
(4) °K to °C = °K – 273.15
(5) °F to °R = °F + 459.7
(6) °R to °F = °R – 459.7
(7) °C to °R = °C × 4/5
(8) °R to °C = °R × 5/4

7.2 CONCLUSION

Members of the shooting community throughout the world should no longer experience a difficulty in converting a temperature scale from one to another. They are quite simple to work out correctly.

CHAPTER 8

BULLET GEOMETRY

CONTENTS

"***Geometry*** *(from the Ancient Greek: γεωμετρία; geo- "earth," -metron "measurement") is a branch of mathematics concerned with questions of shape, size, relative position of figures, and the properties of space.*"

—Wikipedia – the free encyclopedia

Frequently, it has been necessary for the serious bullet maker and ammunition reloader in the shooting community to calculate the physical or geometric attributes of a bullet. With varying configurations – shapes and sizes – even the calculation of surface areas can be tedious, among other things.

We have five basic shapes to make up and form into the shape of a bullet:

(1) Round-nose to shape the bullet's head;
(2) Spitzer points to shape the bullet's head;
(3) Cylinder to form the sides of the bullet; and
(4) Flat surfaces to shape either the bullet's base or its flat head.

We have six basic bullet types:

(1) Wad cutter (WC);
(2) Semi-Wad cutter (SWC);

(3) Round-Nose (RN);
(4) Flat-Nose (FN);
(5) Spitzer Point (SP); and
(6) Spitzer Point Boat-Tail (SPBT).

8.1 THE FIVE BASIC SHAPES

If we were to use the following five basic shapes, we could then assemble the shape of a bullet and, with these shapes, calculate their physical attributes.

Round Nose

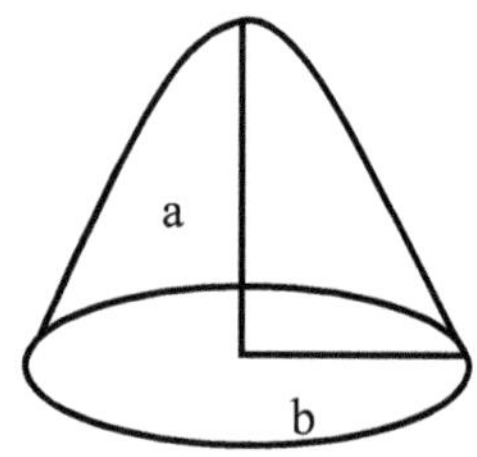

$$\text{Volume} = \frac{1}{2}\pi b^2 a$$
$$\text{Surface area} = \pi b(b+a)$$

Spitzer Nose

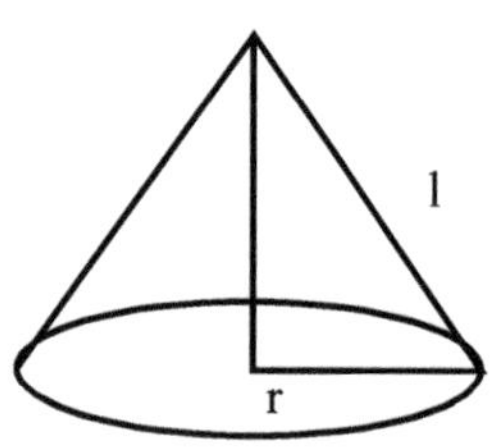

$$\text{Volume} = \frac{1}{3} r^2 h$$
$$\text{Surface area} = \pi r \sqrt{r^2} + h^2 = \pi r l$$

Bullet Cylinder

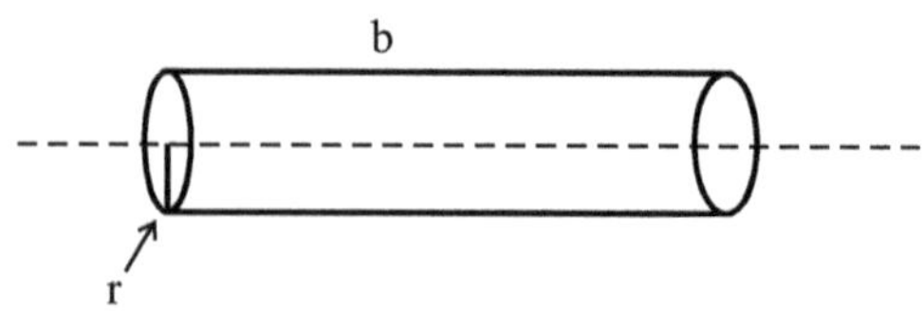

$$\text{Volume} = \pi r^2 h$$
$$\text{Surface} = 2\pi r h$$

Round Ball

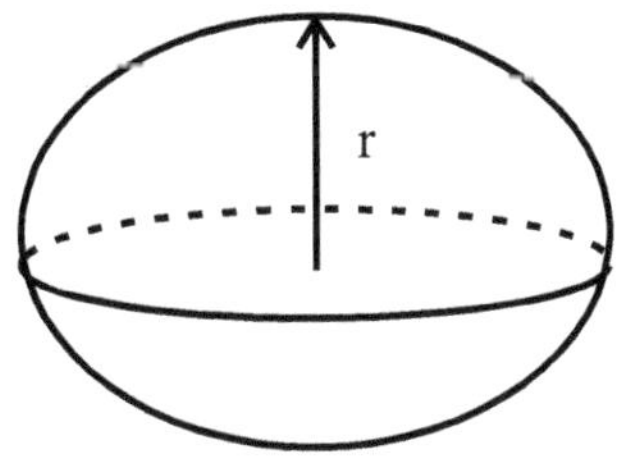

$$\text{Volume} = \frac{4}{3}\pi r^2$$
$$\text{Surface area} = 4\pi r^2$$

Base

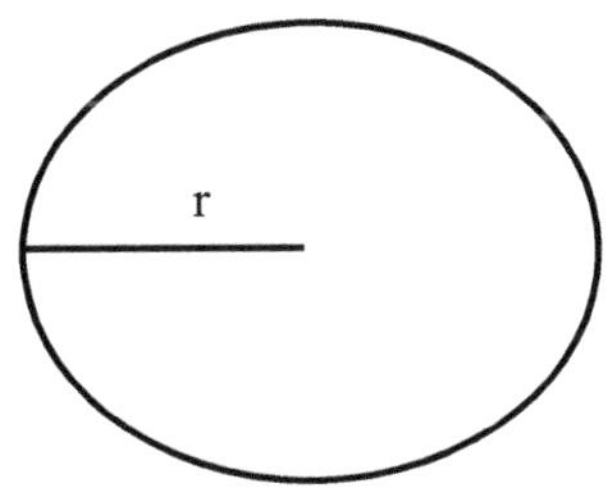

$$\text{Surface area} = \pi r^2$$

8.2 THE SIX BASIC BULLET TYPES

The following six basic bullet types, made up of the preceding five basic shapes, contain the following mathematical equations in order to calculate their entire physical attributes.

Spitzer Bullet w/Flat Base

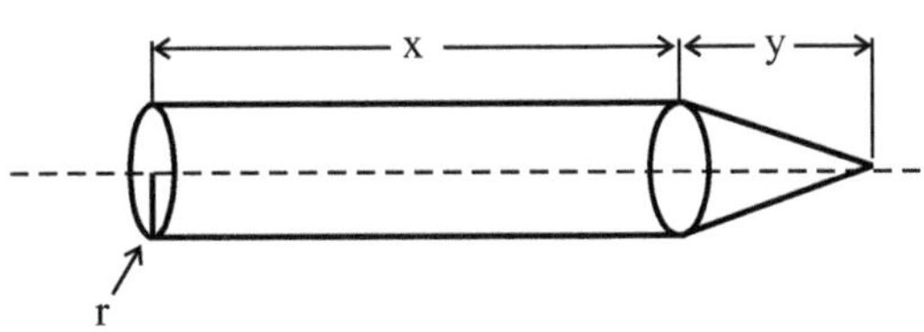

$$\text{Volume} = (\pi r^2 x) + (\pi r^2 y/3)$$
$$\text{Area} = (2\pi r x) + (\pi r y)$$
$$\text{Surface area} = (2\pi r)(r+x) + (\pi y)(r+y)$$
$$\text{Base} = \pi r^2$$

Spitzer Bullet w/Boat-Tail

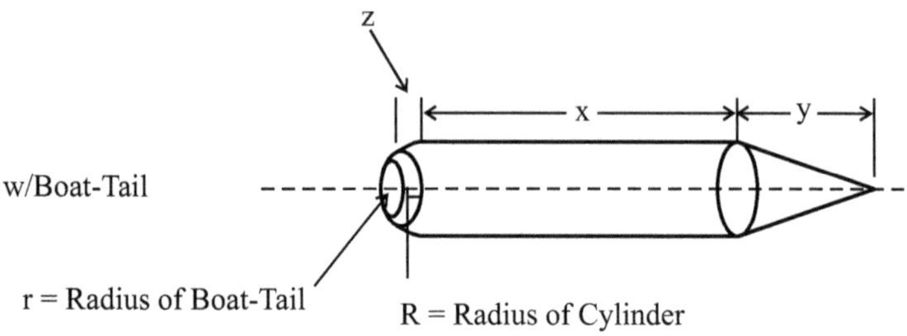

$$\text{Volume} = (\pi R^2 x) + (\pi R^2 y/3)\ [(\pi y/3)\ (R^2{+}R{+}r{+}r^2)]$$
$$\text{Area} = (2\pi Rx) + [\pi(R{+}r)z] + (\pi Ry)$$
$$\text{Surface area} = [(2\pi R)(R{+}x)] + \pi[R^2 + (R{+}r^2)(z{+}r^2)] + \pi R(R{+}y)$$
$$\text{Base} = \pi r^2$$

Flat-Nose Bullet w/Flat Base

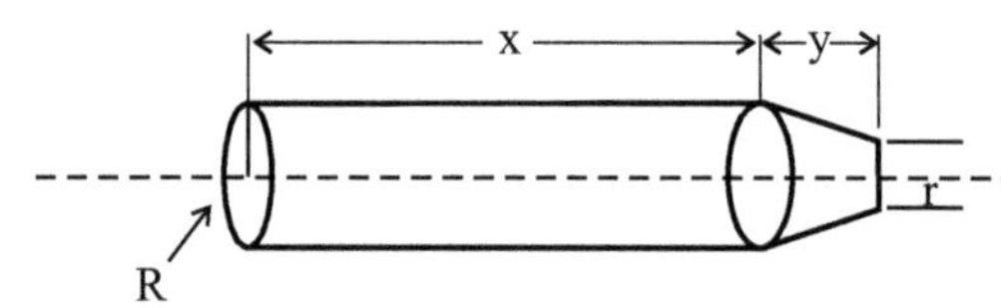

$$\text{Volume} = (\pi R^2 x) + [(\pi y/3)(R^2{+}R{+}r{+}r^2)]$$
$$\text{Area} = (2\pi Rx) + [\pi(R{+}r)y]$$
$$\text{Surface area} = [(2\pi R)(R{+}x)] + \pi[R^2{+}(R{+}r^2)(y{+}r^2)]$$
$$\text{Base} = \pi R^2$$

CHAPTER 9

STATISTICS

CONTENTS

"The history of statistics can be said to start around 1749 although, over time, there have been changes to the interpretation of the word statistics. In early times, the meaning was restricted to information about states. This was later extended to include all collections of information of all types, and later still it was extended to include the analysis and interpretation of such data. In modern terms, "statistics" means both sets of collected information, as in national accounts and temperature records, and analytical work which require statistical inference."

—Wikipedia, the free encyclopedia

9.1 INTRODUCTION

Statistics represents a special branch of the science of mathematics that deals with the collection, organization, and analysis of numerical data to allow the scientist, the technician, the engineer, or any ordinary person, with a need to know for that matter, to interpret and use information correctly for the explicit purpose of making competent decisions.

It has been the principle concern of most members of the shooting community to examine the numerical data responsible for shooting accuracy. If it does not concern or contribute to accuracy, or the variables affecting accuracy, then most members simply lack the interest to study it. Truthfully, if the numerical data prove irrelevant to the science of small arms ballistics, or the variables responsible for accuracy, it has no real relevance to us anyway.

Nor is it necessary for us to delve in the advance or "higher statistics" of the normal numerical data we commonly encounter in our collective quest for knowledge about guns and ballistics. Only the expert scientist or engineer finds it truly necessary to venture beyond the use of basic statistical tools of collecting, organizing, and analyzing ballistic numerical data. To the average member of the shooting community, these basic statistical tools are entirely adequate most of the time.

9.2 BASIC STATISTICAL TOOLS

For us, as members of the shooting community, but not necessarily including or excluding the academic community, there are four basic statistical tools we can consistently find useful in either the treatment or interpretation of ballistic numerical data:

(1) mean average;
(2) standard deviation;
(3) extreme spread;
(4) probability.

9.2.1 MEAN AVERAGE

"*In mathematics and statistics ... is simply the mean or **average** when the context is clear ... the sum of a collection of numbers divided by the number of numbers in the collection.*" – Wikipedia, the free encyclopedia
The measurement of mean average as a statistical tool, such as the mean average of muzzle velocity in the distribution of 100 chronographic readings, represents an accurate and useful measurement of the central

tendency of a particular lot of ammunition, with a particular gun and at a particular period and place in time. It is a useful datum point for us to anticipate the muzzle velocity of a particular lot of ammunition.

$$\text{Mean average} = \frac{\sum^{i=1} \chi_i \ldots\ldots \chi\infty}{N} \tag{1}$$

As the equation says, the summation (Σ) of χ_i to $\chi\infty$, with χ_i representing a numerical datum point, such as muzzle velocity, and $\chi\infty$ representing an infinite number of datum points, divides by "*N*," the number of scores, measurements, or observations, such as the number of muzzle velocity readings from an electronic chronograph, represents the mean average or central tendency of this distribution of scores, measurements or observations.

9.2.2 STANDARD DEVIATION

"*In statistics, the standard deviation (Sd) (represented by the Greek letter sigma,* **σ**) *is a measure that is used to quantify the amount of variation or dispersion of a set of data values. A low standard deviation indicates that the data points tend to be very close to the mean (also called the expected value) of the set, while a high standard deviation indicates that the data points are spread out over a wider range of values.*" – Wikipedia, the free encyclopedia

Traditionally, standard deviation has been defined as the square root of the square of the sums ($\Sigma\chi^2$) of a distribution of scores divided by the number of scores (*N*) – minus the sum of the squares ($\Sigma\chi^2$) – and then divided again by the same number of scores (*N*); or:

$$\text{Sd} = \frac{\sqrt{\Sigma\chi^2 - (\Sigma\chi)^2 / N}}{N} \tag{2}$$

Standard deviation serves as a statistical tool for us to judge the pattern of dispersion from the central tendency, such as the pattern of dispersion from the average muzzle velocity in a distribution of muzzle velocities (scores). It helps us to interpret a pattern in the distribution of scores and to apply our interpretation more meaningfully.

9.2.3 EXTREME SPREAD

"In a firing accuracy test, the distance between the two shots farthest from each other."

—The Free Dictionary by Farlex

Extreme spread is a simple concept to understand. It is the difference between the lowest score and the highest score in a distribution of scores, and helps us to locate and to identify the range of frequencies in a given distribution.

$$ES = d\chi \tag{3}$$

9.2.4 PROBABILITY

"Probability theory is the branch of mathematics concerned with probability, the analysis of random phenomena. The central objects of probability theory are random variables, stochastic processes, and events: mathematical abstractions of non-deterministic events or measured quantities that may either be single occurrences or evolve over time in an apparently random fashion. If an individual coin toss or the roll of dice is considered to be a random event, then if repeated many times, the sequence of random events will exhibit certain patterns, which can be studied and predicted. Two representative mathematical results describing such patterns are the law of large numbers and the central limit theorem."

—Wikipedia, the free encyclopedia

Probability, in the context of the science of ballistics for the shooting community, would represent at most a statistical tool to help us to predict the next muzzle velocity or anything else – such as the maximum chamber pressure, point of impact downrange, distance from baseline in the calculation of trajectory, depth of penetration, etc. – after we have obtained enough data in order to make a prediction. The greater the amount of data, the greater is the accuracy we can make in our prediction.

$$\text{Probability} = \frac{\Sigma\chi}{N} - \frac{\text{ES}}{2} \tag{4}$$

where $\Sigma\chi$ = summation of the number of scores; N = number of scores; ES = extreme spread.

The above formula (4), a derivative of the formula to calculate the probability in the throw of dice, is a function of the summation of scores, such as a series of muzzle velocity readings in a test or experiment, divided by the number of scores, or readings, minus one half of the extreme spread (ES) of those scores.

9.3 CONCLUSION

Statistics deals with the collection, organization, and analysis of numerical data allowing us to interpret and use information more correctly and, perhaps, more precisely for the purpose of helping us to make more intelligent decisions.

For us in the shooting community, there are four basic statistical tools we can use consistently in our quest for accuracy:

(1) mean average;
(2) standard deviation;
(3) extreme spread; and
(4) probability.

Mean average is the central tendency of frequencies in the distribution of scores.

Standard deviation is a statistical tool to judge the pattern of dispersion relative to the central tendency.

Extreme spread represents the difference between the lowest and the highest scores in a distribution of scores.

Probability helps us to predict the next score, such as muzzle velocity, once we have gathered enough data for such a prediction.

When we use these statistical tools properly, with the proper numerical data, we immediately put ourselves in a position to make more intelligent and competent interpretations, judgments, and decisions.

SECTION TWO

CHAPTER 10

THE SCIENCE OF EXTERIOR BALLISTICS

CONTENTS

"External ballistics or exterior ballistics is the part of ballistics that deals with the behavior of a non-powered projectile in flight. External ballistics is frequently associated with firearms, and deals with the unpowered free-flight phase of the bullet after it exits the barrel and before it hits the target, so it lies between transitional ballistics and ballistics. However, external ballistics is also concerned with the free-flight of rockets and other projectiles, such as balls, arrows, etc."

—Wikipedia, the free encyclopedia

10.1 DEFINITION

Exterior ballistics is the scientific study of the patterns and relationships of the effects and characteristics of the physical environment over the free flight characteristics of the bullet.

Its study starts at the precise moment the bullet leaves the barrel, continues throughout the bullet's entire free flight path, and finally stops at the precise moment the bullet strikes the target but before it begins to penetrate or transfer its kinetic energy into it.

It includes the calculations and predictions of a bullet's trajectory and all physical attributes that affect its flight characteristics.

It also includes the effects and characteristics of our planet's gravitational forces, rotational movements, and its atmospheric envelope, including the effects of the Sun, the Moon, and all the effects and characteristics of the gravitational and electromagnetic forces and movements of the solar system that affect a bullet in free flight.

CHAPTER 11

THE FIELD-EFFECT THEORY

CONTENTS

11.1 DEFINITION

"*In physics, the **Field Effect** refers to the modulation of the electrical conductivity of a material by the application of an external electric field.*"

—Wikipedia, the free encyclopedia

In the science of ballistics, "'*field-effect' refers to the modulation of sunlight and gravity over a bullet's flight path.*"

—Alvah Buckmore, Jr.

11.2 INTRODUCTION

One day back in the hot and humid summer of 1979, while shooting at silhouettes in the Westfield Sportsman Club (Massachusetts), we stopped shooting in order to observe another man shooting at a metallic target on a pair of metal rails.

Finding his experimental work most interesting, we walked over to this man's table to observe his work more closely. On his table, we found some neatly written data representing the experiment's results. As we dropped down our head to read the written material on the table, the man said, "Oh! Please don't pay any attention to that material … I can't get consistent results …"

On the table was a gun mounted in a bench-rest and a pair of chronographic screens about 15 feet from the end of the barrel.

Twenty-five feet from the table was the target directly in front of the chronographic screens. It was a piece of metal one inch thick, four inches wide, and four inches high mounted on a rail-road track mechanism of about four feet long and with the individual rails about 3½ inches apart from each other.

Alongside the target and mechanism was a common yardstick in the English unit of measurement. The man would place the metal target at the front of the rails. Then, every time a bullet would hit the metal target, the target would withdraw backwards due to the release and transfer of kinetic energy from the bullet to the target. After that, he would walk back down the range to measure the distance of travel with his yardstick and to move the target back to its original position on the rails for another shot.

In spite of his efforts, he complained, he could not get his loads to shoot consistently from shot to shot. He said, while pointing to his data on the table, "The same loads shoot differently every day." Some days, such as yesterday, he noted, the same loads knocked the metal target further on the rails than today. He contributed this problem to his loads. They were too imprecisely reloaded to shoot consistently, he thought. "I need better reloading equipment," he was truly bewildered and frustrated, he admitted.

Thinking, we looked down at the data; looked across to the chronographic screens; looked downrange at the target; and, finally, looked up to the sky and saw the Sun almost directly overhead.

Looking down again, we read the date, time, and temperature of the data entries.

Looking up at the sky again, without saying a word to our friend, we began to perceive and recognize relationships, never before noticed that day, between velocity, kinetic energy, momentum, recoil, air density, time in the direction of sun-light, and to the relative position of the Sun in the sky.

11.3 THEORY

In a scientifically valid experiment, if we were to fire a gun at one-hour intervals, starting from seven o'clock in the morning until seven o'clock the following morning, at the same or similar target, we would eventually verify the field-effect theory.

At seven o'clock in the morning, with the temperature cooler and air denser than later in the afternoon, the bullet striking the metallic target would cause it to move backwards on its tracks for a given, easily measurable distance.

Conversely, as the Sun climbs higher and higher over the horizon, the air in the bullet's flight path becomes less and less dense with each incremental increase in temperature. Bullet velocity increases as the air density decreases; then kinetic energy increases with the increase in velocity.

Because velocity and kinetic energy increases with each incremental increase in temperature, as the day progresses, the target moves further and further away with each successive shot.

Our friend, unaware of the effects and characteristics of sunlight over air density in the bullet's flight path, had falsely assumed his "problem" to have been caused by faulty ammunition. In his naivety, he mistakenly assumed a given combination of primer, gunpowder, and bullet weight would yield a *given and constant velocity* – apparently 24 hours a day! He was wrong!

11.4 THE SOURCE OF ENERGY

We have two sources of energy in this field-effect phenomenon: solar conduction (Sc) and solar convection (Sv).

Solar conduction is the direct transmission of solar energy from the Sun.

Solar convection is the re-transmission of solar energy from the Earth's surface. Depending on soil and water properties, their physical properties and capacity to store or reflect solar energy from the Sun, the surface of the Earth will absorb some of the Sun's energy during the daytime and then release most of it during the nighttime. At the same time, during the daytime, with the Sun somewhere in between the two horizons, while absorbing solar energy, the Earth will simultaneously release some of it as well.

This means, among other things, the air density in the bullet's flight path will either warm up or cool off directly due to energy from the Sun and (reflecting or releasing) energy from the Earth's surface.

So, the mathematical relationship between solar conduction and solar convection to field-effect is the following:

$$\Xi = \sum_{MV \to \xi}^{n=\infty} (\Delta Sc, \Delta Sv) \tag{1}$$

where field-effect (Ξ) is the summation (Σ) of each increment (Δ) of solar conduction (*Sc*) and each increment (Δ) of solar convection (*Sv*) in the bullet's flight path from the muzzle velocity (*MV*) to the initial terminal velocity (ξ). The number of computations (n) is infinite (∞).

The unit of measurement is either calories or British thermal units (BTU).

Though the number of increments for computation can be infinite ($n = \infty$) in the bullet's flight path, the computations start from MV and stop at the precise moment the bullet strikes the target at ξ, but before it begins penetration or transfers energy into it.

11.5 THE FIELD-EFFECT CHARACTERISTIC CURVE

As reported earlier, as the day progresses from seven o'clock in the morning to mid-day, we will observe a progression of travel on the railroad tracks after each successive shot in which velocity, kinetic energy, momentum, and recoil also increase successively. Noise level decreases as the day progresses due to the progressive reduction of air density, however.

As our Sun approaches the horizon, from mid-day to dust, each successive shot thereafter causes a successive reduction in the distance of travel on the railroad tracks.

Drawing a characteristic curve on a piece of graph paper (Figure 11.1), the curve rises rapidly, starting from seven o'clock in the morning to mid-day – its peak and then rapidly falls to a very low level when the Sun drops below the horizon. From then on, the characteristic curve *nearly flattens out* with only a gradual decline throughout early and late evening; as early morning arrives, just before the Sun, still below the horizon, starts to warm the air, the characteristic curve rises with the rising Sun. When the Sun pops over the horizon, the characteristic curve rises more rapidly after each successive shot until it reaches a position of numerical value, on the graph paper, similar to the position it may have held 24 hours earlier.

11.6 REAL-TIME STUDY

Naturally, in a real-time study, the actual field-effect is much more complex than the way we have just described it. As an example, cloudcover may obscure direct observation of the Sun – either partially or completely over the bullet's flight path – momentarily or continuously – while the total effect between S_c and S_v will correspondingly increase or decrease the summation of values in each increment of the bullet's flight path, with sometimes solar convection more important than solar conduction in day-time. Ordinarily, it is the other way around.

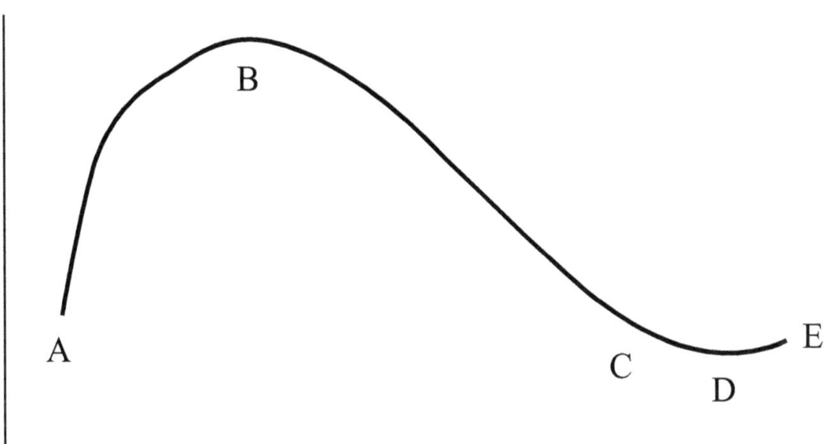

FIGURE 11.1 The field-effect characteristic curve.

11.7 PROGRESSION OF CURVES

As we know very well, the Earth rotates around the Sun in an elliptical path and takes a year to complete a full cycle. As the distance and angle of incident (the angle at which the sunlight strikes the Earth) is different each day, the amount of energy striking the Earth's surface is different, too, with the amount of energy gradually *decreasing* with approaching winter or gradually *increasing* with approaching summer.

This phenomenon accounts for the Progression of Curves (Figure 11.2) in the field-effect theory. It means that each 24-hour characteristic curve (Figure 11.1) will be slightly different than the preceding 24-hour characteristic curve. With approaching summer, this curve will rise in relatively uniform steps of progression, with slight variations each day depending on a wide variety of technical and climatic conditions such as clouds, storms, rain, snow, relative humidity, temperature, and cooling and warming winds. With approaching winter, of course, the opposite effect occurs (Figure 11.2).

Any typical 24-hour characteristic curve will consist of several basic components necessary for us to fully understand the field-effect theory.

- A–E is the daily curve;
- B is the peak of the daily curve for a given day;
- C–D is the nightly curve;

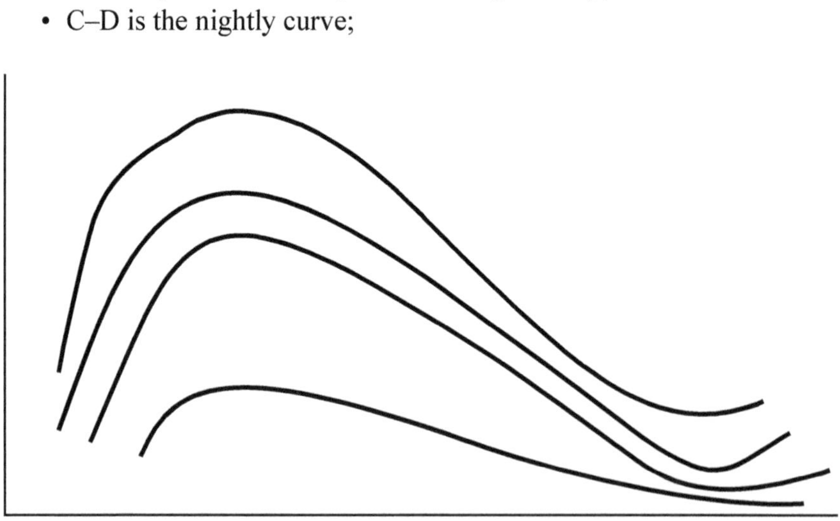

FIGURE 11.2 Progression of curves of the field-effect theory.

- C – E is the early morning transitional curve and the carrying threshold important to the successful operation of automatic weapon systems.

In the summertime, this *Carrying Threshold* is probably unimportant to automatic weapons most of the time; however, in the winter – particularly in cold winters – the *Carrying Threshold* may seriously alter the time-pressure curve (Figure 1.4) and burning-rate characteristics of gunpowders and hence the operation of the automatic action and cyclic rate of fire. So, it is critically important, particularly in military applications, to design, develop, and employ a gunpowder that will function properly in all carrying thresholds in the progression of curves throughout the year in any climatic system. Sometimes, it may prove necessary to change the gunpowder, to a faster or slower burning one, when entering a very cold or very hot climatic region (sometimes along with a "hotter" or "colder" primer).

In very cold regions, the operation of the automatic action may become too sluggish to be militarily useful and reliable. On the other hand, the cyclic rate of fire in a very hot climate may be far too fast for the design of the automatic action to be safe for the gun, and certainly may cause an unnecessary expenditure of ammunition, and may have been the case in Vietnam along with a tendency for the soldiers to fire far too many rounds due to its mild recoil and the panic they felt during combat.

In any event, throughout any position of the 24-hour characteristic curve, the operation of the automatic action and its subsequent cyclic rate of fire will vary with the characteristic curve, even with additives in the gunpowder to reduce its sensitivity to temperature (and, in part, due to the lubricants' and the buffer spring's sensitivity to temperature).

In view of the progression of curves, the cyclic rate of fire of any automatic action will vary each day, with a tendency to *increase* with approaching summer and to *decrease* with approaching winter.

In addition, the buffer springs, rebound slides, the main springs and tension bars will vary in both tension and "feel" with the progression of curves, from day to day with revolvers and, like automatic weapon systems, possess a tendency to *increase* tension with approaching summer and *decrease* with approaching winter.

Sometimes, when too light but operative in the summertime, a main spring or tension bar may not operate entirely or consistently in the wintertime.

When we adjust the tension on the main spring and rebound slide of a revolver during the summertime, we must remember to be careful to avoid making them too light. Perhaps, in view of these circumstances, it is better to adjust them during the wintertime.

If very light in the summertime, even when very pleasant to shoot, with just 1½ to 2½ pounds of pressure to pull back the trigger (for serious target work and competition), instead of the 5 to 15 pounds of pressure some firearms manufactures mandate in their handguns (for self-defense and to avoid liability in a court of law), they may not operate the action satisfactorily during the wintertime or during a cold evening or early morning. Nor may the hammer strike the primer with sufficient force to cause an ignition.

In fact, the serious shooter, who does his own honing and stoning of the trigger assembly, frequently finds it necessary to adjust revolver springs twice a year, once for the winter and once for the summer. Firearm manufactures, of course, install springs much stronger than really necessary to ensure complete reliability throughout the entire year of the progression of curves, perhaps without their realizing they are dealing with the progression of curves. They may just want to make the action safe to use for everyone regardless of his measure or level of competency.

CHAPTER 12

A THEORY OF THE EFFECT OF FIELD-EFFECT OVER TIME

CONTENTS

12.1 INTRODUCTION

If we were to spend some time out of doors during a very late evening or very, very early morning, it will become perceptible to most of us the amount of time between the times on a clock is longer than the same amount of time during a warmer period of time in the daytime.

If we were to get up every morning at the same time, day-in-day-out, every day of the year – to brush our teeth, to comb our hair, to wash and shave our face and to eat our breakfast – it will soon become very apparent the amount of time it takes to perform these simple but important tasks vary considerably every day.

When we wind our wristwatch every morning, equally as perceptible, in the middle of winter, the mainspring takes longer to wind than it does in the middle of summer.

As each day approaches summer, though the distance between the intervals or graduated lines of time on the clock stays constant, unless something distorts their distances, the duration of time between those intervals or graduated lines shrink and, as each day approaches winter, the duration of those intervals or graduated lines expand.

When we perform the same tasks every day at the same time, the amount of time it takes to perform these same tasks will increase as the temperature increases and decrease as the temperature decreases.

As we stated earlier, the field-effect affects the air's density in the bullet's flight path. When temperature increases, due to an increase in the amount of solar energy from the Sun, the time-pressure curve and burning rate characteristics of the gunpowder changes to produce a higher chamber pressure, a shorter time-pressure curve, and a faster burning powder, and hence a higher muzzle velocity and a loss of timing as the gunpowder stops burning before the bullet leaves the barrel. Then, the air density in the bullet's flight path also decreases which increases the bullet's trajectory velocity, kinetic energy, and momentum, due to a drop in aerodynamic drag on the bullet. A higher muzzle velocity and trajectory velocity will give us a flatter trajectory (see The Effect of Field-Effect Over Trajectory).

Likewise, when temperature decreases, due to a decrease in the amount of solar energy from the Sun, the time-pressure curve and burning rate characteristics of the gunpowder changes to produce a lower chamber pressure, a longer time-pressure curve, and a slower burning powder, and hence a lower muzzle velocity and a loss of timing as the gunpowder continues to burn a little after the bullet leaves the barrel.

Then, the air density in the bullet's flight path also increases which decreases the bullet's trajectory velocity, kinetic energy, and momentum, due to an increase in aerodynamic drag on the bullet. A lower muzzle velocity and trajectory velocity will give us a more parabolic trajectory

At the same time, however, there is an additional variable responsible to affect the bullet's trajectory – time! The amount of solar energy in a bullet's flight path increases as the Sun rises; then time contracts and time of flight takes more time. The amount of solar energy decreases with the setting Sun; then time expands and time of flight takes less time.

So, along with a reduction of the air density in the bullet's flight path, which gives us a higher velocity and flatter trajectory, when temperature increases due to an increase in solar energy from the Sun, the amount of time it takes for the bullet to complete its trajectory lengthens. When the temperature decreases, time expands and the bullet takes less time to complete its trajectory and to become more parabolic due to the corresponding reduction in trajectory velocity from an increase in aerodynamic drag.

Therefore, the mathematical relationship of the effect of field-effect (Ξ') over time (t) is the product of field-effect (Ξ) and time (t) to time (T').

$$T' = \Xi \times t \tag{1}$$

where Ξ = field-effect is the summation of solar conduction and solar convection in each increment of air in the bullet's flight path; t = actual time measured by a clock; T = effect of field effect over time.

12.2 FURTHER CLARIFICATION

To clarify this concept of the expansion and contraction of time, think of time as a straight line which we must break down into uniform, constant, and equidistant intervals (Figure 12.1), the way we customarily think of time.

Then, if something were to modulate time – cause it to expand or contract – though the number of intervals would stay constant, the distance between them would increase with the expansion of time and decrease with its contraction (Figure 12.2).

Or think of time as a ruler with uniform lines of graduation at 1/64-inch intervals. Make it from a material highly sensitive to temperature. When the temperature is 70°F, the distances between the intervals are correct at 1/64 of an inch; however, as the temperature decreases, the ruler stretches uniformly throughout its entire length, which would automatically increase the distance between each interval.

If the ruler were not a ruler but time, and if something were to cause it to expand, then the amount of time it would take to perform a certain task would decrease relative to the new distance between the intervals.

This phenomenon of stretching time would, in effect, shorten time or the amount of time it would take to perform a given task.

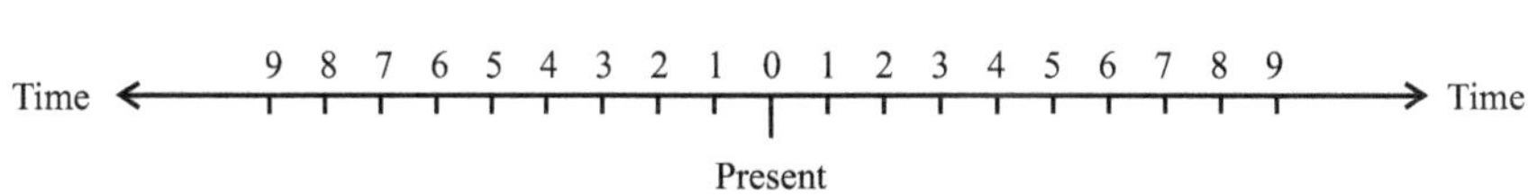

FIGURE 12.1

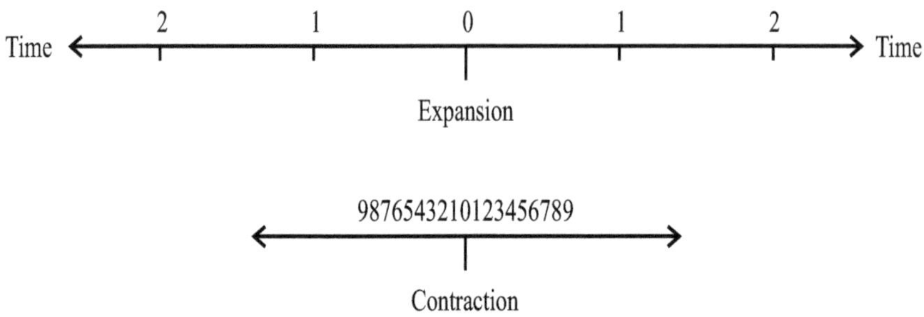

FIGURE 12.2

In the science of small arms ballistics, the amount of time it would take a bullet to travel through its trajectory would shorten when time expands and increase when time contracts. In effect, it would appear as if the bullet's velocity were decreasing with an expansion of time and increasing with its contraction.

As the illustrations in Figures 12.2 through 12.4 clearly demonstrate, time is longer in the nighttime and in the early morning hours of the day with the Sun still below or slightly over the horizon. During mid-day, after the Sun has had hours to warm the Earth's surface (Figures 12.4) time is the shortest time of the day. Then, as the Sun approaches the opposite

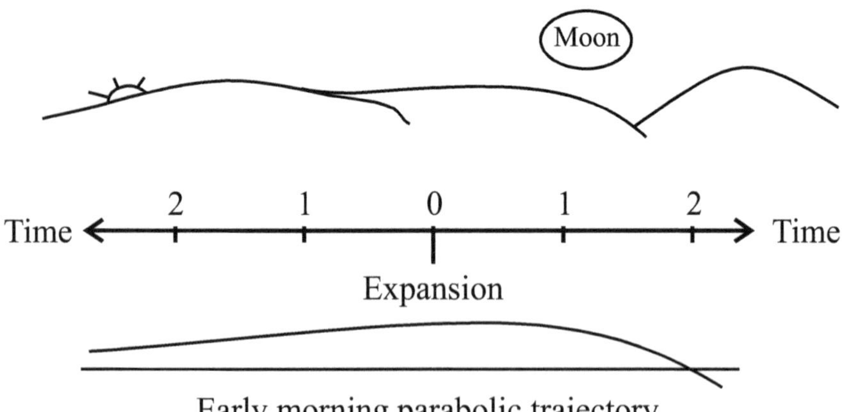

FIGURE 12.3

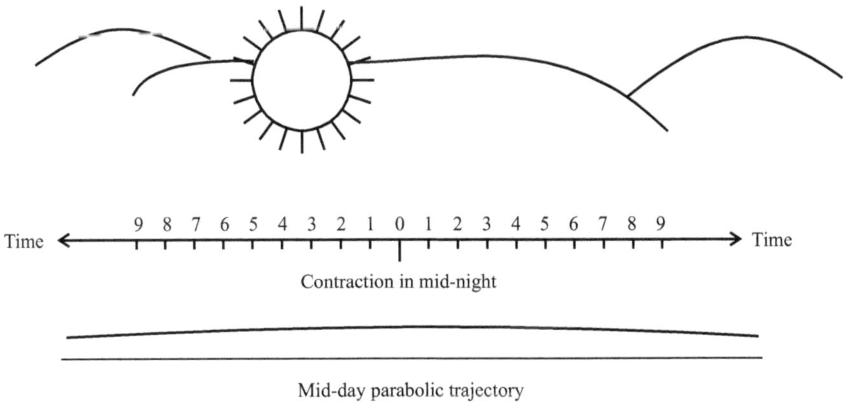

FIGURE 12.4

horizon, the day slowly cools down and the time plasma begins to expand again to separate the graduated lines for longer time duration during the nighttime (Figures 12.5).

To repeat, in the nighttime and the early morning hours, the distances between each interval of time is the longest (Figures 12.4 and 12.5); as the day warms from the Sun climbing over the horizon, time contracts and the distances between each time interval shortens (Figure 12.3). As the day cools, the time again expands as the distanccs between each interval

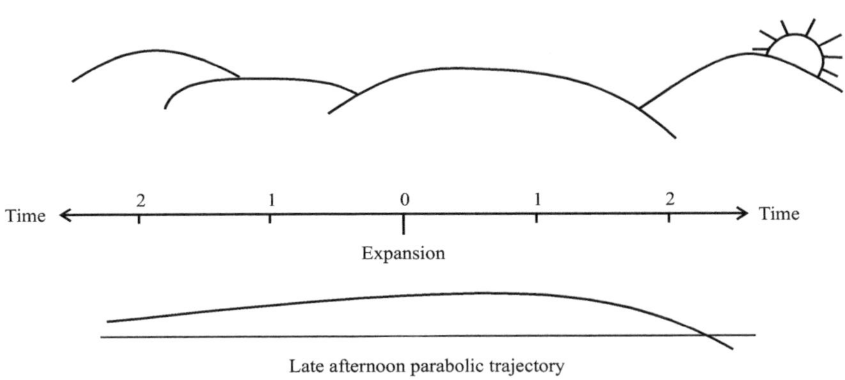

FIGURE 12.5

lengthens (Figure 12.4). So, the effect of field-effect over time has the same characteristic curve as the field-effect characteristic curve (Figure 1.4) (see The Field-Effect theory), including the progression of curves.

Interestingly, as time contracts or expands constantly throughout any 24-hour day, it not only affects the bullet's time of flight but its trajectory as well (see A Theory of the Effect of Field-Effect Over Trajectory).

12.3 CONCLUSION

In our preliminary analysis of this subject, it gradually became obvious there was a growing possibility for the development of another science, possibly several new sciences, along with new technologies and industries evolving from the science of small arms ballistics.

When we look at field-effect and the effect of field-effect over a bullet's trajectory and time itself, it becomes almost flagrantly obvious, that, in order for the Sun and sunlight to modulate time – to cause it to expand or contract over a period of 24 hours every day – means *time has to be a physical entity*. It can no longer be a simple abstraction, concept or a convenient way to measure equidistant intervals in a 24-hour day from a clock, or even a simple variable or subexpression in the formula to calculate speed when we know distance. We now know that distances and velocities change with the modulation of time and that time varies with solar energy and gravity (see A Theory of the Effect of Gravity Over Time). It is not a constant!

Time, for the Sun and sunlight, later, when we will discover gravity as well, to modulate its properties – to make it expand or contract – must be a physical substance, most likely a plasma, and the only place we can find such a plasma comes from the Sun itself. Indeed, our whole Universe contains this plasma. It is everywhere! It is this *star stuff emanating from our Sun and other stars that make up the stuff of time* and, when we research this *star stuff* in outer space, we will soon discover that it is indeed a plasma and contains a small electrostatic charge to it as well, which may explain the reason light travels faster in outer space than in a vacuum. It may be that it is this small electrostatic charge in outer space that is responsible for modulating the velocity of light.

When we examine a bullet traveling down its flight path, in addition to the three components of drag – drag induced by the transfer of energy, drag induced by the surface area of the nose, and drag induced by the surface area of the cylinder – there is also a more unpredictable but a calculable variable *drag or accelerate* induced by the field-effect's modulation over time and trajectory.

Now we go into the effect of gravity over time, distance, direction, and velocity.

CHAPTER 13

A THEORY OF THE EFFECT OF GRAVITY OVER TIME

CONTENTS

13.1 INTRODUCTION

"Gravitation or gravity is a natural phenomenon by which all physical bodies attract each other. Gravity gives weight to physical objects and causes them to fall toward one another."

—Wikipedia, the free encyclopedia

In Chapter 12 – A Theory of the Effect of Field-Effect Over Time – we demonstrated a relationship between solar energy from the Sun to the expansion and contraction of time. As the Sun slides from the eastern sky to the western sky, time – in a relative position of expansion before sun-rise – slowly contracts with the rising Sun to reduce the distance between the intervals of time and therefore increases time. When the Sun moves beyond mid-day, the warmest time of the day, time – at its greatest period of contraction for the day – begins to expand. Time then stretches out to separate the distance between the graduated lines of time to make time longer.

But there is another variable in the contraction and expansion of time – gravity. Gravity has the same effect over time, and relationship to time, as field-effect.

Mathematically, the effect of gravity (g') is a relationship of the product of gravity (g) and time (t) to time (T").

$$T" = g \times t \tag{1}$$

where g = The Sun's gravitational pull or any other gravitational pull from elsewhere; t = actual time by a clock; T" = effect of gravity over time.

13.2 CONCLUSION

Because the effect of gravity over time has some of the same effects as the effect of field-effect over time, it has the same mathematical relationship.

When we shoot a gun, gravity constantly pulls the bullet down toward the ground and its acceleration, or the speed it falls to the ground, depends on the latitudinal lines on the planet Earth. From either the North or South Pole, as we approach the Equator, the speed gets slower and slower until it reaches it minimum speed at the Equator of 32.0862 ft/s^2. That effect and its acceleration affect the bullet's trajectory velocity, kinetic energy, and momentum, transfer of energy, the maximum effective range, the maximum range of lethality, the maximum distance downrange and the depth of penetration in a given target.

Gravity's effect over time has the effect of modulating time through compression. With the planet Earth in a constant state of rotation around its axis, this modulation of the time plasma around its circumference has a rotary motion pattern to it and an oscillation with a vector relationship when we combine field-effect's modulation with gravity's modulation, meaning both magnitude from gravity and direction from solar energy.

In addition, while the Earth's gravitational pull compresses the time plasma around its circumference, the Moon with its gravitational pull has the opposite effect of decompressing the time plasma and the Sun, with a far superior gravitational pull, causes a much stronger decompression toward its direction.

With the Earth rotating around its axis, the Moon rotating around the Earth, and both of them rotating around the Sun, there is a constant moving ridge or swell on the surface of the planet consisting of a compressing/

decompressing magnitude of the plasma with the Sun's solar energy of expanding/contracting the plasma in a horizontal direction.

With the Sun's solar effect over the direction of the time plasma and the three sources of gravity's (Earth, Moon, and Sun) effect over its magnitude, we have a complex vector relationship of magnitude and direction.

We must remember, when dealing with precision calculations, the effects of gravity over time and time of flight are important to consider, particularly when dealing with great distances, ordinarily not a serious problem in small arm ballistics.

CHAPTER 14

A THEORY OF THE EFFECT OF FIELD-EFFECT OVER TRAJECTORY

CONTENTS

14.1 INTRODUCTION

If we were to chronograph the muzzle velocity of a gun early in the morning – say, at seven o'clock – and again at two o'clock in the afternoon, the muzzle velocity of the gun will be different each and every time, and most likely a little higher the second time at two o'clock, depending on the weather.

This phenomenon is due directly to the effect of the Sun warming the air around the gun's receiver and barrel. As the ambient temperature elevates with a corresponding increase in temperatures inside of the chamber and barrel before and after ignition, the gunpowder changes some of its burning characteristics, namely the amount of time it takes to burn. When it increases its burning speed, the time-pressure curve shortens – all of which may increase the muzzle velocity, depending on the gunpowder and the length of the barrel. If the gunpowder should burn too fast, due to the increase in temperatures inside and outside of the gun, it may stop burning long before the bullet leaves the barrel, giving a slightly lower muzzle velocity and putting the bullet out of time as it leaves the barrel. Remember! The bullet must leave the barrel at the precise moment the gunpowder stops burning for optimum accuracy, and the rapid changes in temperatures can easily knock it out of time.

Then, because the Sun also warms the air in the bullet's flight path, the air density decreases with each incremental increase in temperature, producing both a higher and flatter trajectory velocity.

With a lower muzzle velocity at seven o'clock in the morning, there will be of course a lower trajectory velocity and therefore a more parabolic trajectory curve. At any given distance downrange, the bullet will also impact higher on the target.

Then, at two o'clock in the afternoon, with a higher temperature, the muzzle velocity and trajectory velocity will be higher. This increase in both velocities, relative to the velocities at seven o'clock, will produce a flatter trajectory. Impact on the target downrange will be a little lower.

Because the gun is also relatively "cold" on its first shot, it will shoot even lower and slightly to the right of center in the Northern Hemisphere, depending on the distance downrange, though we may not notice it except at very long distances (see the coriolis effect), and to the left in the Southern Hemisphere. As the barrel warms with each successive shot, it will take several shots for the barrel's temperature to stabilize and finally for us to locate its new shooting position (see A Theory of the Asymptotic Function), usually and probably slightly just below the shooting position of seven o'clock.

14.2 CONCLUSION

Naturally, in a *real-time* study of our physical environment, with a constant change in field-effect and the Effect of Field-Effect Over Trajectory, there will be constant and subtle changes in temperatures, damp air densities, and barometric pressures – from the gun to the entire bullet's flight path.

So, muzzle velocity and trajectory velocity will go up and down constantly throughout the day (see Figure 14.1); on the most part, however, unless these changes become dramatic and significant, we need not worry about them. It only becomes important and significant when we must shoot for extreme precision at long distances (most of us need not

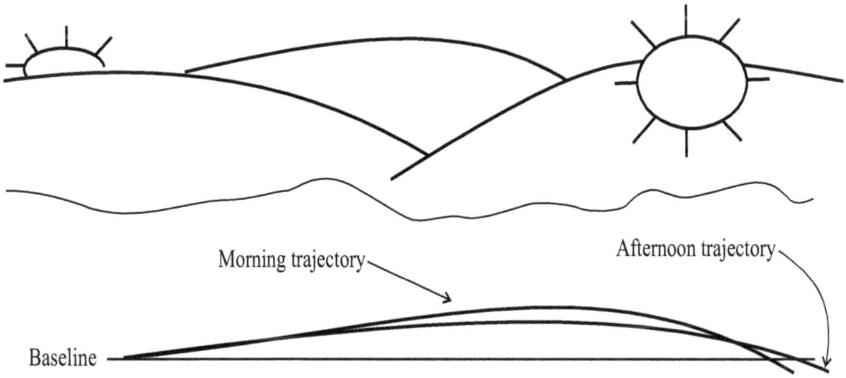

FIGURE 14.1.

worry) starting from early morning and working through the afternoon. Moreover, the coriolis effect also becomes critically important in these calculations.

CHAPTER 15

THEORY, APPLICATION, AND CALCULATION OF TRAJECTORY IN *REAL-TIME*

CONTENTS

"In physics, the ballistic trajectory of a projectile is the path that a thrown or launched projectile will take under the action of gravity, neglecting all other forces, such as friction from air resistance, without propulsion. The United States Department of Defense and NATO define a ballistic trajectory as a trajectory traced after the propulsive force is terminated and the body is acted upon only by gravity and aerodynamic drag."

—Wikipedia, the free encyclopedia

"The firearm was originally invented in China during the 13th century AD, after the Chinese invented gunpowder during the 9th century AD. These inventions were later transmitted to the Middle East, Africa, and Europe. The world's first firearm in history was the fire lance, the prototype of the gun. The fire lance was invented in China during the 10th century and it is the predecessor of all firearms."

—Wikipedia, the free encyclopedia

In the beginning of guns and ballistics – sometime between the 13th and the 14th centuries, but well after the first "Fire Lance" in the 10th century China – before we had finally realized bullets (or projectiles) travel in a parabolic trajectory; we held the assumption, not aware of the effects of the aerodynamic drags, gravity, and solar energy; that bullets travel in a straight line; reach a maximum distance for a given gun and powder-charge, and then abruptly drop to the ground – also in a straight line.

Later, when some ballisticians, gunners, and military ordnance specialists had begun to perceive the existence of a parabolic curve and flight path, the inconsistencies and contradictions of the earlier assumptions of those two straight lines compelled further research – the research which incidentally continues today.

We were slow to recognize the effects of gravity and its relationships to the bullet's mass, momentum, kinetic energy, and velocity – as well as to its trajectory.

In the 1830s, there was a recognition of the "coriolis effect" (see coriolis effect), the tendency of bullets to deflect to the right of the target in the Northern Hemisphere and to the left in the Southern Hemisphere.

Between 1977 and 1979, during a series of experiments and observations at the Westfield Sportsman's Club, we began to perceive field-effect and it effect over a bullet's flight path.

Field-effect deals with the effects of sunlight over the density of air, time, time of flight, and the three components of drag induced by the transfer of energy from the bullet to each increment of the bullet's penetration in its flight path; drag induced by the surface area of the nose and drag induced by the surface area of the cylinder. The base does not create drag; the trueness of its roundness in its manufacturing process, however, determines the design limitations of its accuracy in flight. Analogous to automobile tires, the greater the roundness of the tires, the greater is the mileage we can obtain within the design limitations of its material.

Once ballisticians began to perceive the concept of parabolic flight and trajectory, there was almost a continuous struggle, from a variety of sources all over the world, in several scientific disciplines, to develop methods and mathematical formulas to calculate trajectory within a reasonable accuracy. In the end, there were some genuinely number-crunching computations that only a computer – at first an analog computer before and during World War

II, and then a digital computer – could perform satisfactorily save for field artillery, ships and coastal guns. Microcomputers came much later.

For small arm ballistics, those methods were cumbersome and painfully unrealistic. In no way could we run around with a several hundred pound analog computer on top of a rifle or handgun. Until recently, even the thought of mounting a ballistic computer on a rifle was out of the question. The technology was just not available! Now it is a "piece of cake."

To be successful, we needed to know the bullet's "ballistic coefficient" ["*a measure of its ability to overcome air resistance in flight*"– Wikipedia, the free encyclopedia]; the height of the sight above the center of the bore; the bullet's muzzle velocity and all of its velocities at each interval of computation (usually at every 100 yards or 100 meters); drop of the bullet from the line of departure from the center of the bore; bullet's time of flight at each interval; bullet's path from line of sight; distance of each interval; distance from muzzle to sighting "zero;" drop from bore at a range equal to the sighting "zero;" and drop from bore at the target.

These computations were useful only at relatively short distances, namely just a few hundred yards. With some exceptions, these computations were based only on 100 yard (or 100 meter) intervals. At the first 100 yards, the computations were rarely more than 90% accurate; however, by the time they got out to 500 yards, their accuracy would be rarely more than 60%. With a 40% drop in accuracy, it is agonizingly obvious we would need a much better system of computing trajectory. To get anywhere near 90% accuracy at 500 yards, we would need to compute – not in intervals – but in increments of cubic inches, or in other equally suitable units of measurement. With 3,600 increments of cubic inches for every 100 yards, that means 3,600 separate sets of computations or 18,000 separate sets of computations out to 500 yards. Unless we have an exceptional capability in mental mathematics, we would need to use a high-speed microcomputer.

15.1 THE THREE COMPONENTS OF DRAG

In Figure 15.1, we can clearly illustrate the first component of drag – transfer of energy. Transfer of energy (*TE*) is the product of kinetic energy (*KE*) and air density (σ) where

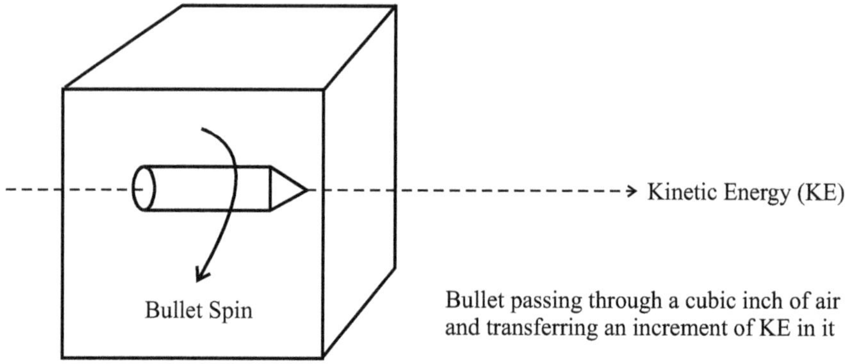

FIGURE 15.1

$$TE = KE \times \sigma \tag{1}$$

When we increase the bullet's velocity, we increase the bullet's kinetic energy and therefore the amount of kinetic energy it can transfer into every increment of the bullet's flight path and then into the target when it reaches it.

If we increase the air density in the bullet's flight path, on the other hand, we increase the transfer of energy going into each increment of air, from the bullet, and therefore reduce the bullet's remaining velocity and kinetic energy as it travels into the next increment of penetration.

If the air density decreases, though, principally due to an increase of ambient temperature, the transfer of energy decreases with a corresponding increase in the bullet's remaining velocity and kinetic energy as it travels into the next increment of penetration and, eventually, the target.

Of course, during periods of little sunlight, the winter, or during the nighttime, the air density would have to be much greater than in bright sunlight of the daytime to cause a greater transfer of kinetic energy into each increment of penetration in the bullet's flight path. This increase in air density would have to affect the bullet's trajectory velocity, as well as the muzzle velocity, to reduce the amount of kinetic energy available when it strikes the target.

We need to know the remaining bullet velocity at each interval downrange in order to calculate its flight path and trajectory. That requires us to identify and calculate every component of drag from the muzzle to each interval. As stated above, the first component of drag is the transfer of energy.

We need to calculate that transfer of energy from the bullet into each increment of penetration before we can go any further.

This method is the Little Method (LTTE) and actually a function of integrated calculus:

$$LTTE = \lim_{MV \to \xi}^{n=\infty} \int \Sigma\, [p\Delta KE - \Delta KE \text{ x } \Delta\sigma)] \qquad (2)$$

where LTTE = the loss of kinetic energy in the bullet through the transfer of energy in each increment of penetration in the bullet's flight path. *p*Δ*KE* = the preceding increment of kinetic energy in the bullet's flight. Δ*KE* = the kinetic energy available in the increment of penetration.

LTTE means the "Loss through Transfer of Energy." Take the first increment of penetration immediately in front of the muzzle to calculate the transfer of energy by multiplying the first increment of air density by the bullet's kinetic energy at the muzzle. Once done, then go to the second increment of penetration, and so on, until you will have completed all computations to the first interval of 100 yards (or 100 meters).

Then, you can subtract that answer, which will be very small, from the preceding increment of kinetic energy to get the loss of kinetic energy. Do this 3,600 times for every interval of 100 yards to get the remaining velocity in the bullet's trajectory. But, to be most accurate, you must also include the other two components of drag.

Nose drag compression (Np) is the second component of drag and represents the drag induced by the surface area of the nose.

$$Np = \frac{\sqrt{SA^2 \text{ x } V^2}}{K} \qquad (3)$$

where *Np* = drag compression of the nose surface area (lb/cu./in.); *SA* = surface area of the bullet's nose (sq.in.); *V* = bullet's forward velocity (fps) in a given increment of penetration in the bullet's flight path. $K = 1,964,636.542$ for bullets 100 grains or greater in weight and 19,646,365.42 for bullets less than 100 grains in weight.

As the bullet moves downrange, it compresses the air in front of the nose and around its cylinder.

So, the degree or amount of compression depends directly on the surface area of the nose directly in front of the bullet – for any given value of velocity. The greater the surface area, the greater is the compression we will find in front of the bullet.

When we increase velocity, we increase drag compression. It is this drag compression in front of the nose responsible for helping to slow down the bullet in each increment of it flight path – from increment to increment.

It has nothing to do with the transfer of energy or the velocity of the transfer of energy. There is always transfer of energy; however, the rate of the bullet's deceleration depends on the actual configuration of the bullet, but namely, its frontal and cylinder surface areas.

Actual configuration will not affect transfer of energy, either; this is because the transfer of energy is the product of the bullet's kinetic energy and the air's density in the flight path, not drag compression.

Nevertheless, bullet configuration will most definitely affect the rate of deceleration through the two drag compressions.

Once we compute the value of transfer of energy, we proceed to calculate the effect of the nose's frontal surface area over drag compression, in order to add its value to the value of the transfer of energy, so that we can determine the total drag responsible for the rate of deceleration.

But we have one more major component of drag to add in our calculation of the rate of deceleration and the value of velocity at any given increment or interval downrange in order to determine the bullet's remaining lethality.

Cylinder drag compression (Cp) is the third and last component of bullet drag. It has the same effect over the rate of deceleration as nose drag compression with one single exception – the amount of the effect.

$$\text{Cp} = \frac{3\sqrt{\text{SA}^2 \text{ x } \text{V}^2}}{\text{K}} \tag{3}$$

15.2 TOTAL DRAG COMPRESSION

Total drag compression usually begins with a large difference between the effect of Np over the effect of Cp by an enormous 90% with bullets greater than 100 grains in weight and 68% with bullets less than 100 grains in weight. Cp is always substantially greater than Np.

However, as the bullet moves downrange, this ratio slowly increases with each successive increment of penetration into its flight path with 97% for bullets greater than 100 grains in weight and 85.3% for bullets less than 100 grains in weight – at 500 yards.

With bullets greater than 100 grains in weight, Cp is approximately 11 times greater than Np; as the bullets move downrange to 500 yards, with Np staying almost constantly the same value with each increment of penetration, the Cp constantly increases with each successive increment of penetration until it becomes in excess of 34.5 times greater than Np.

With bullets less than 100 grains in weight, however, Cp is approximately 3.4 times greater than Np at the muzzle, and gradually increases to a value of 6.8 times greater than Np at 500 yards.

Drag compression, due to the surface area of the nose (Np) and the surface area of the cylinder (Cp), rapidly drops with the drop in velocity at each increment in the bullet's flight path.

With bullets 100 grains or more in weight, the difference between Np and Cp at the muzzle is approximately 69.5% and drops down to 45.4% at 500 yards.

While, on the other hand, the difference between Np and Cp with bullets less than 100 grains in weight, at the muzzle, is 59.7% and drops down to 21.5% at 500 yards.

Apparently, the ratio between Np and Cp, as it grows successively smaller with each increment of penetration, increases the cylinder drag compression, which grows progressively greater, in respect to the nose compression, and begins to cause a more significant influence at lower velocities, particularly below the speed of sound.

We believe, once the bullet drops below the speed of sound (approximately 1130 fps at sea level), if it has a boat-tail configuration, the effect of this configuration is to cause a significant reduction in the cylinder drag compression, effectuating the effect of reducing the length and therefore the total surface areas of the cylinder and cylinder drag compression. This phenomenon of a boat-tail configuration seems to have the effect of causing the bullet to lose velocity at an even slower rate once it will have dropped below the speed of sound; this is the reason behind the common use of boat-tail configurations among competitors and in the military services.

15.3 PRELIMINARY CONCLUSION

Larger bullets with greater length, diameter, and weight, for any given velocity, lose velocity at a slower rate than smaller bullets with shorter lengths, smaller diameters, and lesser weights due to the ratio between nose drag compression and cylinder drag compression, and probably due to the ratio of the cylinder's mass to the mass of the bullet's nose.

A boat-tail configuration seems to have the effect of reducing the total surface area of the bullet's cylinder and therefore the effect of reducing cylinder drag compression when the bullet drops below the speed of sound.

When a bullet leaves the barrel, it automatically goes into a state of deceleration, principally due to the transfer of energy from the bullet to the first increment of penetration in its flight path.

Then, the actual configuration of the bullet determines the rate of deceleration, which successively diminishes with each successive drop in velocity.

Whether a bullet of a .17 caliber, leaving the barrel at a remarkable velocity of 4,100 fps, or a .50 caliber bullet leaving the barrel at 2,750 fps, it cannot reach or pass beyond 600 yards before it will have run out of lethality if we were to believe Major General Julian Hatcher (June 26, 1888–December 4, 1963) in his "Hatcher's Notebook".

According to his studies on the science of ballistics between World War I and World War II, he drew the conclusion that it takes about 60 ft/lbs of kinetic energy to effectively kill a man. Kinetic energy is the lethality in the bullet that kills or causes destruction to our bodies or property. Our studies have determined that no bullet between the .17 caliber and the .50 caliber can survive pass 600 yards and still maintain that level of lethality of 60 ft/lbs. Yet, there are literally hundreds of documented accounts of military snipers killing enemy soldiers well beyond 600 yards. Obviously, 60 ft/lbs is not the real threshold of lethality and/or, apparently, it depends on the actual circumstances, such as field-effect in the bullet's flight path causing the bullet to travel substantially beyond 600 yards to carry more than 60 ft/lbs of kinetic energy.

Beyond that magic range of 60 ft/lbs at 600 yards, the bullet may continue to travel for a considerable distance and still carry enough lethality to effectively kill a man.

In fact, it may travel for miles if we were to raise the barrel to a significant angle, say 20 or 40 degrees; however, once it loses its forward momentum due to the combination of transfer of energy, nose drag compression and cylinder drag compression, it will simply drop in a long curve, even possibly tumble in flight, if light enough, to eventually fall to the Earth miles away from the original shooting site. If it should hit anyone, it will lack sufficient kinetic energy, usually a small fraction of a ft/lb, to be lethal but still may sting a little, particularly on the face, in the way a BB may sting.

If, even now with a forward velocity, but well below the range of lethality, as General Hatcher defined it, a bullet could continue to penetrate living tissue to cause bloody and painful damage.

More than 80 years ago, he mounted a .30 caliber machinegun on a stable platform with the muzzle pointing straight up in the air. With a pair of binoculars and several miles of open space, later water in a small pond, while operating in an underground bunker, he repeatedly fired the machinegun and waited for the bullets to strike the ground or water. He reported feeling deep frustration in his powerlessness to spot or find them striking the ground or water. Apparently, the bullets would reach a maximum elevation several hundred yards up in the air to follow a long shallow curve to his right, pursuant to the Earth's gravitational pull and the Correlis Effect, to strike the Earth many, many miles away from his shooting site.

Earth's rotation and the Correlis Effect will guarantee that the bullets will not fall anywhere near his shooting site and, most likely without special instrumentation, prove almost impossible for him to locate – putting to death a lot of ancient theories.

15.4 A COMPUTER ALGORITHM TO CALCULATE LTTE

What follows is the mathematical algorithm to support a computer program to calculate the loss of both the bullet's kinetic energy and velocity through transfer of energy in each increment of penetration in the bullet's flight path, including loss through nose drag and cylinder drag compression with an automatic compensation and correction for a given bullet weight. It was written originally in 1990 using the Basic language.

ALGORITHM 1

```
100 Print "Loss through Transfer of Energy with automatic correction
    of bullet weight"
110 Print: Input "Velocity at the muzzle = ___," V
120 Print: Input "Bullet's weight = ___," W
130 Print: Input "Distance in inches = ____," Inches
140 DEFDBL P, S: CLS
150 K = ((W/225218) * V^2) / 2
160 T = ((W/225218) * V^2) / 2
170 E = T-(T * .0000509)
180 A = 1964636.542^ (1/2)
190 H = 1964636.542^ (1/2)
200 DEFDBL J, S
210 IF W < 100 THEN GOTO 370
220 FOR I =1 TO INCHES
230 Y = SQR ((2*K) / (W/225218))
240 J = (H/(Y^ (1/2) /.0000509 ^ (1/2))) ^2
250 S = (A/(Y^ (1/2) /.0000509 ^ (1/2))) ^2
260 P = ((S * ((SQR ((2 * K) / (W / 225218))))) / 1964636.542)
270 Z = ((J * ((SQR ((2 * K) / (W / 225218))))) / 19646365.42)
280 O = E * P + Z
290 K = T – O: T = K
300 O = E * P + Z
310 K = T – O
```

```
320 LOCATE 12, 35: PRINT I
330 NEXT I
340 LOCATE 12: PRINT I, Y, K, P
350 PRINT CHR$(7)
360 END
370 FOR I – 1 TO INCHES
380 Y = SQR ((2 * K) / (W / 225218))
390 S = (A / (Y ^ (1/2) / .0000509 ^ (1 /2))) * 3
400 P = ((S ^2 * Y ^2) ^ (1/3)) / 1964636.542
410 J = (A / (Y ^ (1/2) / .0000509 ^ (1/2))) * 3
420 Z = ((J ^ 2 * Y ^2) ^ (1/3)) / 19646365.42
430 O = E * P + Z
440 K = T – O: T = K
450 O = E * P + Z
460 K = T – 0: T = K
470 O = E * P + Z
480 K = T – O
490 LOCATE 12, 35: PRINT I
500 NEXT I
510 LOCATE 12: PRINT I, Y, K, P
520 PRINT CHR$(7): END
530 END
```

The above computer algorithm and program will not calculate, compute, or measure a bullet's trajectory; but it will provide the critical steps toward the calculation and prediction of trajectory with additional algorithms. LTTE is an important tool to calculate the loss of velocity and kinetic energy at every increment of the bullet's penetration in its flight path. Once we can calculate and predict the remaining velocity at any given interval, starting at 100 yards, etc., we can then either draw a trajectory curve on a piece of draft paper or display trajectory almost in *real-time* on a computer screen.

$$D = \left[2\sqrt{Vf / MV}\right](64.32 \times T^2) \tag{4}$$

where D = Drop from bore (inches); MV = Muzzle Velocity (fps); Vf = Velocity at the first interval (such as 100 yards) in fps; T = Time of flight (seconds).

When a bullet leaves the bore, it immediately drops a little relative to the center of the bore. The above equation calculates that drop.

Before we can go any further with our calculations, we must calculate the bullet's time of flight to the first interval, such as 100 yards.

$$T = \frac{R}{(MV + Vf)/2} \tag{5}$$

where R = Distance from muzzle to target (feet); MV = Muzzle Velocity (fps); Vf = Velocity at the first interval (fps).

Then, with the above calculation completed, we can continue to calculate the height of the bullet from the baseline – an imaginary line starting from the center of the bore to the center of the target. Because a bullet's trajectory or flight path tends to be either above or below the baseline, we need to know only its exact position at each interval (100, 200, 300 yards, etc.) to draw an accurate line of trajectory, representing the bullet's flight path, from interval to interval on a sheet of paper. Or, with today's computer technology and the available software, we can easily program a computer to display a line of trajectory on its screen. To complete those calculations, however, we must use the following equation:

In the following equation, we calculate the time of flight from muzzle to each interval (100 yards, etc.) in order to calculate the drop from the center of the bore to each interval.

$$Pb = \left[\frac{R}{Rs}(Rs + H)\right] - (Df + H) \tag{6}$$

where Pb = Bullet path from line of sight (inches); R = Distance from muzzle to target (feet); Rs = Distance from muzzle to sighting zero (feet); H = Height of scope (or open sight) from center of the sight or scope to the center of the bore (inches); Df = Drop from bore at the target (inches).

Then, we calculate the position of the bullet at each interval, either above or below the baseline. When we finish all calculations at each interval and compile all data, we draw a line of trajectory on a sheet of graph paper or program a computer to do the work for us.

We will find a computer infinitely easier and much, much faster. While in near *real-time*, a computer today can instantly display a growing line of trajectory as it computes the data points at each increment and calculates the bullet drop at each interval.

When we first started to develop this method of computation in the late 1980s, it took almost 30 minutes with the Radio Shack TRS-80 pocket computer to calculate LTTE out to 100 yards without the additional computations of the other two components of drag. Within a few months of constant usage, we successfully burned out that pocket computer.

The Tandy 2000 microcomputer took almost 20 minutes to do the same thing using MS-DOS 2.11 as the operating system with an 80186 processor at about 8 MHz.

When the 80386 microcomputer became available, we increased our speed of computations to about 9 minutes for every 100 yards.

With the 80486 33MHz microcomputer, the speed of computations jumped to approximately 9 seconds for every 100 yards.

CHAPTER 16

WIND DEFLECTION

CONTENTS

Wind deflection – the deflection of a projectile resulting from the effects of wind."

—The Free Dictionary by Farlex

After taking into consideration all the variables responsible for accuracy, we must still include the calculation of wind deflection in order to hit the target downrange – particularly at extreme long distances.

Wind deflection (Dw) takes the following mathematical expression:

$$Dw = Vw\,[t - (R/MV)] \tag{1}$$

where Dw = wind deflection (inches); Vw = velocity of 90° cross-wind (inches per second); t = time of flight (seconds); MV = muzzle velocity (fps); R = distance downrange from muzzle to target (feet).

KEYWORDS

- **cross-wind**
- **muzzle velocity**
- **wind deflection**

CHAPTER 17

AIR DENSITY IN *REAL-TIME*

"The density of air, ρ (Greek: rho) (air density), is the mass per unit volume of Earth's atmosphere. Air density, like air pressure, decreases with increasing altitude. It also changes with variation in temperature or humidity. At sea level and at 15°C, air has a density of approximately 1.225 kg/m³ (0.001225 g/cm³, 0.0023769 slug/ft³, 0.0765 lbm/ft³) according to ISA (International Standard Atmosphere).

The air density is a property used in many branches of science as aeronautics; gravimetric analysis; the air-conditioning industry; atmospheric research and meteorology; the agricultural engineering in their modeling and tracking of Soil-Vegetation-Atmosphere-Transfer (SVAT) models; and the engineering community that deals with compressed air from industry utility, heating, dry and cooling process in industry like a cooling towers, vacuum and deep vacuum processes, high pressure processes, the gas and light oil combustion processes that power our turbine-powered airplanes, gas turbine-powered generators and heating furnaces, and air conditioning from deep mines to space capsules."

—Wikipedia, the free encyclopedia

In the calculation of trajectory in *real-time*, as discovered in the chapter on "Theory, Application and calculation of Trajectory in *real-time*," we used the "Little Method" (LTTE) – based on the relationship of transfer of energy to kinetic energy and air density – to calculate trajectory (with the other two components of drag).

We also discovered the use of ordinary dry air density to be entirely adequate, and surprisingly accurate, most of time in most situations, even at moderately long distances.

However, as we have learned in the chapter on The Effect of Field-Effect Over Trajectory, we must use damp air density in *real-time* to accurately calculate field-effect in *real-time.*

$$\sigma d \; \frac{\dfrac{70.56\left[29.92\left(e^{-hy}\right)\right]}{53.3\left(460+T^{\circ}F\right)}}{1728} \tag{1}$$

where σd = damp air density (lbs/cubic inch); h = constant: 0.00003158; y = elevation above sea level (feet); $T°$ F = temperature (Fahrenheit).

The subexpression, 29.92 (e^{-hy}), represents the equation to calculate the uncorrected barometric pressure (UBP), where

$$\text{UBP} = 29.92\ (e^{-hy})$$

Thus, as stated in an earlier chapter, the equation to calculate damp air density does indeed contain barometric pressure as a variable.

Ordinarily, dry air density of 0.0000509 pounds per cubic inch is more than reasonably accurate in the calculation of a bullet's trajectory at short ranges – say, 300 yards. However, at longer distances, it will become very important to include the calculation of damp air density when dealing with the requirement of extreme precision or working in a very moist environment, such as in the case of rainfall. A moist environment will, of course, counter-intuitively, always provide for a lower air density with a higher velocity in the bullet's flight path at every increment of penetration, and that will change shot placements and produce larger groups.

When we shoot for groups in rain, such as in a competitive event, even at short distances, such as 100 to 300 yards, it will prove a real challenge with most calibers smaller than a .50 caliber. Groups get erratic and unreliable. Sometimes, it would be better for us to shoot at a barn! Now, seriously, ordinarily, shooting at a man or a moose in rain works as long as we do not expect a tight group.

The relationship between mass and volume, with moisture – almost always the case in most environments – will make the air density smaller than dry air density due to the moisture's effect of reducing density; because the molar mass of water (the ratio of the mass of a given substance to the amount of substance) is less than the molar mass of dry air. That automatically reduces density, and that affects a bullet's trajectory. Using dry air density as one of the variables in the calculation of trajectory will ordinarily produce completely negligible errors at short distances. When dealing with a requirement for much greater accuracy and precision, then, of course, we would have to include the value of damp air density, which is normally the common issue in an outside environment.

CHAPTER 18

THE SPEED OF SOUND IN AIR

"The speed of sound is the distance travelled per unit of time by a sound wave propagating through an elastic medium. In dry air at 20°C (68°F), the speed of sound is 343.2 meters per second (1,126 ft/s). This is 1,236 kilometers per hour (667 kn; 768 mph), or a kilometer in 2.914 seconds or a mile in 4.689 seconds."

—Wikipedia, the free encyclopedia

Over these many centuries there have been many, many attempts to measure the velocity of sound, and perhaps, it was Sir Isaac Newton (December 25, 1642–March 20, 1726/7) who, for the first time, successfully computed the speed of sound in air as 979 feet per second, approximately within 15% of the true value. However, he failed to perceive or, if he did, neglected to measure the effect of fluctuating and unpredictable temperatures with the change in elevation and climate. It was Pierre-Simon, marquis de Laplace (March 23, 1749–March 5, 1827) who corrected that problem.

It took awhile before the scientific community began to perceive that sound travels at different velocities in different mediums, whether in a gas (air), a liquid (water) or a solid (concrete).

Throughout the 17th century, there were several significant attempts to measure the speed of sound more accurately. People such Marin Mersenne (September 8, 1588–September 1, 1648) in 1630, Pierre Gassendi (January 22, 1592–October 24, 1655) in 1635, and Robert Boyle (January 25, 1627–December 31, 1691) came much closer than their earlier peers.

Then, in 1709, Reverend William Derham (November 26, 1657–April 5, 1735), the Rector of Upminster, England, published a much more accurate measurement of the speed of sound. His method was unique. He used a telescope in the tower of the church of St. Laurence in Upminster, England, to measure the time it took for a flash from a distant shotgun until he heard the report of that shotgun with a half second pendulum. Then, several more measurements were made using shotguns, shooting from several different local landmarks, including the North Ockendon church, using triangulation to measure the speed.

$$Vs = \sqrt{2400(460 + T°F)} \quad (1)$$

where Vs = speed of sound in air (fps); T°F = temperature of air (°F).

CHAPTER 19

APPROXIMATE TIME OF FLIGHT

"Time of flight (TOF) describes a variety of methods that measure the time that it takes for an object, particle or acoustic, electromagnetic or other wave to travel a distance through a medium. This measurement can be used for a time standard (such as an atomic fountain), as a way to measure velocity or path length through a given medium, or as a way to learn about the particle or medium (such as composition or flow rate). The traveling object may be detected directly (e.g., ion detector in mass spectrometry) or indirectly (e.g., light scattered from an object in laser doppler velocimetry)."

—Wikipedia, the free encyclopedia

In any calculation of the time of flight, we will soon realize that it can only be approximate in its trajectory. This system does not consider the effect of field-effect over time or the effect of field-effect over air density or transfer of energy and, subsequently, as well as all aerodynamic drag compression characteristics in the shapes and surface areas of a bullet.

Nevertheless, it is accurate enough for most of us in the shooting community.

$$t = \frac{Ds}{(MV + \xi)/2} \tag{1}$$

where t = approximate time of flight (seconds); Ds = distance from muzzle to target (feet); MV = muzzle velocity (fps); ξ = initial terminal velocity (fps).

CHAPTER 20

MAXIMUM RANGE OF LETHALITY

"*Lethality or deadliness is how capable something is of causing death. Most often it is used when referring to chemical weapons, biological weapons, or their chemical components. The use of this term denotes the ability of these weapons to kill, but also the possibility that they may not kill. Reasons for the lethality of a weapon to be inconsistent, or expressed by percentage, can be as varied as minimized exposure to the weapon, previous exposure to the weapon minimizing susceptibility, degradation of the weapon over time and/or distance, and incorrect deployment of a multi-component weapon.*"

—Wikipedia, the free encyclopedia

"*Stopping power is the ability of a firearm or other weapon to cause enough ballistic trauma to a target (human or animal) to immediately incapacitate (and thus stop) the target. This contrasts with lethality in that stopping power pertains only to a weapon's ability to incapacitate quickly, regardless of whether death ultimately occurs.*

Stopping power is related to the physical properties of the bullet, but the issue is complicated and not easily studied. Although higher caliber has traditionally been widely associated with higher stopping power, the physics involved are multifactorial, with caliber, muzzle velocity, bullet mass, bullet shape, and bullet material all contributing. Critics contend that the importance of "one-shot stop" statistics is overstated, pointing out that most gun encounters do not involve a "shoot once and see how the target reacts" situation.

Stopping power is usually caused not by the force of the bullet but by the damaging effects of the bullet, which are typically a loss of blood, and with it, blood pressure. This is why in many instances a single gunshot wound (GSW), with slow blood loss, does not stop the victim immediately. More immediate effects can result when a bullet damages parts of the central nervous system, such as the spine or brain, or when hydrostatic shock occurs. The importance (or lack thereof) of hydrostatic shock and of momentum transfer in determining stopping power has long been controversial among gun users. Some have ascribed great importance to hydrostatic shock; some have tried to entirely discount it; the truth is somewhere in between. Not every GSW produces it."

—Wikipedia, the free encyclopedia

An examination of lethal energy or kinetic energy in a bullet in Wikipedia, as well as other sources, will readily reveal an enormous interest on the subject by the military armed services, big-game hunters, and even forensic pathologists. There are even individuals writing letters on blogs asking for a definitive answer to these questions plus a mathematical formula or some other method of analysis for them to take to the field.

The problem with this subject is the proper definition of lethality, in this context of small arm ballistics, together with the enormous complexity of calculating it.

Generally speaking, because it is kinetic energy that represents the lethality in a bullet, the heavier the bullet or the greater the velocity, the greater is the lethality. Then, there is that complexity in the design of the bullet. It can have all the lethality we need; but if we design it incorrectly, it will fail to work properly, which happens all the time.

A little more than 30 years ago, there was an experiment using different weights of different bullets at different velocities with approximately the same level of lethal kinetic energy.

With a pile of wooden pine dowels of various diameters, several recreational shooters set up a test to determine the effectiveness of lethal penetration going through those dowels (Figure 20.1). Starting off with a .22 caliber bullet and working up to a .50 caliber, they shot a round of each caliber into those wooden dowels to see what would happen.

Immediately, the pattern became clear and really intuitive for most of us even before the experiment had started. The lighter 52 grain .22 caliber bullets, the same spitzer bullets we used in the original M-16, would partially penetrate the wooden dowels before reflecting off to another direction to strike another wooden dowel, causing each dowel to split, break, or splinter before the bullet ran out energy and stopped entirely. Its path

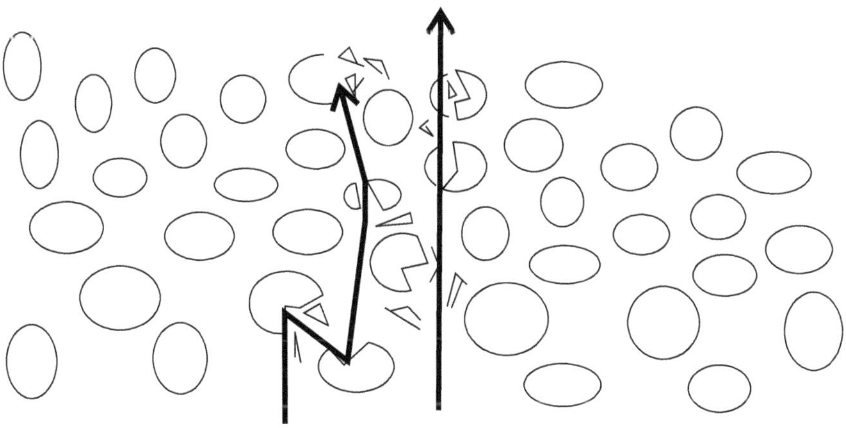

FIGURE 20.1

of penetration was unpredictable and zigzagged through only part of the wooden structure, perhaps following a path of "lease resistance."

When the experimenters continued with the 150 grain .30 caliber bullet, it also followed an unpredictable and zigzagged course but in a straighter line of splitting and splintering dowels before running out of energy a little further down the wooden dowel structure.

Finally, with the heavier 500 grain .50 caliber, the bullet reflected a little to the left and a little to the right of its original path to break, split, and splinter the dowels, but continued in a relatively straight line before penetrating the structure entirely.

The lead core separated from the copper jacket of the .22 caliber bullet, leaving pieces of itself into some of the wooden dowels. Though seriously distorted, that did not happen with the .30 caliber bullet and the .50 caliber bullet rested with only a few scratch marks on the surface area of its nose. All the bullets were full metal jackets.

What we have is clear evidence that lethal kinetic energy is not enough to make a bullet successful in its mission. All bullet weights were well beyond lethality when they struck the wooden dowels. To be successful, we must design and use the correct bullet for a given mission. We can change its weight; change its diameter; change its velocity; alter its configuration with different alloys and jackets; or change its core density from a lead alloy to something such as depleted uranium or tungsten carbide.

Needless to say, we are stuck with a given gun and caliber with severe limitations. Changing a bullet's diameter is not usually an option except, possibly, for the military budgets with the objective to identify, develop, and employ a weapon system to defeat a hardened target such as a tank or bunker. Ordinarily, members of the hunting community are more concerned with the caliber and bullet to defeat a particular game animal, such as something very large and dangerous, than with "hardened targets."

They are particularly concerned with the range of lethality of a given gun and its ammunition. The following algorithm is part of a bigger computer program, originally written in 1990–1991 in Basic, using the 80186 and later the 80286 microcomputer, stored on several 5¼ inch floppy diskettes stored in one of 14 filing cabinets. It is one of many modules and part of a 14,000 line computer program to support a "Spectroanalytic" program on ballistics. We are no longer able to run this program on any of our computers today, because all those formats are incompatible to the present formats in use. Fortunately, we have several hundred pages of "hard copies" of dozens of algorithms, modules, and computer programs we can use as backups for this book but without the ability to operate them from those original formats.

The following Algorithm 2 will calculate, using the three components of drag, the "Maximum Range of Lethality." It is not perfect. However, it will prove surprisingly accurate and reliable in the prediction of any bullet's maximum range of lethality, from the .17 caliber to the .50 caliber, but cannot predict or anticipate the rapid changes in the local climate, the subtle changes in temperature and humidity, or the sudden appearance of a cloud halfway downrange that out of the blue increases the air density, the transfer of energy, the two components of drag compression, the slight reduction in trajectory velocity, or the height of the bullet from the baseline. Nevertheless, it will be remarkably accurate and useful to allow the hunter, particularly, to determine whether he will have enough lethality to effectively kill his game at a given distance. We must remember when dealing with the calculation of such parameters in ballistics, with some important exceptions, the algorithms is always mathematically intensive. Until relatively recently, we did not even have the computing power to make those computations for the consumer.

General Hatcher, as reported in an earlier chapter, said the minimum amount of kinetic energy to effective kill a man was 60 ft/lbs, and NATO reports it as 62.7 ft/lbs. Go to lines 370 and 380. When the program reaches less than 60 ft/lbs of remaining kinetic energy in its calculations, or any value we want, it stops to calculate and display the distance in yards. For large game, such as a moose or bear, 200 ft/lbs seems more realistic, though certainly not definitive, and it is easy to change lines 370 and 380 when calculating the maximum range of lethality for those animals, or for anything we want. From exhaustive research, there does not appear to have been anything done on the minimum amount of kinetic energy to effectively kill wild game.

ALGORITHM 2

```
100 Print : "Maximum Range of Lethality"
110 Print: Input "Muzzle Velocity (fps) = ____," Y
120 Print: Input "Bullet Weight (grains) = ____," W
130 Print: Input "Kinetic energy at the muzzle (ft/lbs) = ____," T
140 Print: Input "Succeeding increment of kinetic energy (ft/lbs) =
    ____," E
150 Count = 0
160 O = E * P
170 K = T – O
180 T = K
190 O = E * P : Y = SQR ((2*K) / W/225218)) : IF W < 100 THEN
    GOTO 260
200 A = 196436.542 ^ (1/2)
210 S = (A/Y^ (1/2) /.0000509 ^ (1/2))) ^2 'NSA DRAG
220 P = SQR ((S^2)* ((SQR ((2*K) / (W/225218))) ^2)) /1964636.542
    'DRAG COMPRESSION
230 H = 19646365.42 ^ (1/2)
240 J = (H/Y ^ (1/2) /.0000509 ^1/2))) ^2 'CSA DRAG COMPRES-
    SION
```

```
250 Z = SQR ((J^2) * ((SQR ((2 * K) / (W/225218))) ^ 2)) / 19646365.42
    : GOTO 320
260 A = (1964636.542) ^ (1/2)
270 S = (A/Y ^ (1/2) / .0000509 ^ (1/2))) * 3 'NSA'DRAG
280 P = ((S ^ 2 * Y ^ 2) ^ (1/3)) / 1964636.542
290 H = 19646365.42 ^ (1/2)
300 J = (H/(Y^ (1/2) / .0000509 ^ (1/2))) * 3
310 Z = ((J ^2 * Y ^ 2) ^ (1/3)) / 19646365.42 'DRAG COMPRES-
    SION
320 K = T – O
330 V = SQR ((2 * K) / (W/225218))
340 PRINT "Remaining Kinetic Energy = ______ (ft/lbs)
350 PRINT K
360 PRINT "Maximum Range of Lethality"
370 IF K > 60 THEN COUNT = COUNT + 1 : GOTO 160 ELSE 380
380 IF K = 60 THEN GOTO 390
390 DEFINT I : I = COUNT / 3600 * 100
400 PRINT "Maximum Range of Lethality = _____ yards"
410 PRINT I : PRINT CHR$(7)
420 END
```

Note: *To obtain "succeeding increment of kinetic energy," multiply the kinetic energy at the muzzle by .0000509 (dry air density) [TE = KE x σ] to get the TE and then subtract that answer from the KE at the muzzle. From there, you will get the remaining KE in the succeeding increment of penetration of the bullet's flight path.*

CHAPTER 21

MAXIMUM EFFECTIVE RANGE

"The effective range (maximum effective range) of a weapon is the furthest distance an effective shot can be taken with reasonable certainty that it will hit. It is determined by a number of factors: type of cartridge fired, inherent precision of the weapon, and volume of fire delivered."

—Wikipedia, the free encyclopedia

- Absolute maximum effective range: "This round is not considered lethal after crossing this threshold" distance. Neither of the other two common "maximum range" values will be greater than this. Purportedly, NATO defines this as the point at which the projectile's kinetic energy dips below 85 joules (62.7 foot-pounds). This is typically claimed when recounting that the P90's effective range is 400 meters on unarmored targets, as classified by NATO. It's worth noting that while the P90 looks neater than the civilian PS90, the extra barrel length increases the muzzle velocity, and thus, the civilian model actually has a longer absolute maximum effective range.
- Maximum effective range on a point target: This is the maximum range at which an average shooter can hit a human-sized target 50% of the time. "Point target" is basically a euphemism for hitting a human torso-sized area in this context. If this range were greater than the absolute maximum, the absolute maximum would be quoted (a non-lethal hit may be accurate, but it's not effective).

—Gun/Wiki

The maximum effective range is …"*The maximum distance at which a weapon may be expected to be accurate and achieve the desired effect.*"
—The Free Dictionary by Farlex

It does not take much thought, research on the Internet, or elsewhere to realize there is no definitive answer to the question: "What is Maximum Effective Range?" In some sources, such as Wikia, the maximum effective range is the "… *further distance an effective shot can be taken with reasonable certainty that it will hit* …" the intended target.

In The Free Dictionary by Farlex, the maximum effective range is "… *at which a weapon may be expected to be accurate and achieve the desire effect.*"

Obviously, we have a problem, i.e., the problem of defining the maximum effective range of a particular gun and its ammunition.

In this thesis on the science of small arms ballistics, we shall define t*he maximum effective range as the greatest distance a particular gun and its ammunition can hold onto a group.*

Discounting the differences in the proficiency of the shooter and the psychological anticipation of felt recoil, plus the sheer physical and psychological fatigue of a prolonged period of shooting a gun with heavy recoil, recoil jump, and loud reports, we will confine ourselves to the mechanics of the gun, its quality in workmanship, and the ballistics of the ammunition, including the consistency of the burning characteristics of the ignition system – the primer; the consistency of the dimensional characteristics of the cartridge case; the quality and consistency of the gunpowder and, of course, the quality and consistency of the bullet's construction. Hitting a target is no big deal. Hitting it consistently, shot after shot, and holding a group at the same time is something entirely different – and much more difficult.

One thing is certain: With all those variables, both physical and psychological, we would need a more definitive and decisive method of analysis. All those variables make it virtually impossible to calculate and predict the maximum effective range with any degree of reliability and consistency. In fact, such a figure would prove meaningless in most instances.

Then, upon a closer examination of this issue, the answer to our question becomes immediately obvious: – Because a bullet can only maintain its stability in flight as long as it continues to fly about the speed of sound, *the threshold of instability is the moment its velocity drops below the speed of sound.* Then, the *groups begin to lose their coherence*, the heavier bullets much more slowly than the lighter bullets.

It is also obvious, when we think it out in detail, many or most bullets, particularly when we fire them from handguns and some rifles, are inherently unable to maintain stability within a few feet of leaving the barrel, even with the correct rate of twist for the bullet at the correct muzzle velocity when it leaves the barrel below the speed of sound.

Experience has taught us the heavier bullets, for any trajectory velocity, lose velocity much more slowly than the lighter bullets. A good example would be the .22 caliber Short. In handguns, it starts off well below the speed of sound and quickly loses momentum to make the maximum effective range not much more than 50 yards and, by the time it reaches 100 yards, maintaining a decent group will prove very difficult even for the best marksman. Firing the same bullet in a high-quality rife would prove only slightly superior beyond 50 yards but not much more effective at 100 yards than with a handgun.

When we use a much heavier bullet (240 grains versus 40 grains), such as the .44 Special from a handgun at 850 fps, though still substantially below the speed of sound when it leaves the barrel, will easily provide superiority over the .22 caliber Short. With groups no different at 50 yards, tests after tests with the .44 caliber will provide much tighter groups at 100 yards than any .22 caliber bullet starting off with the same or greater muzzle velocity (but below the speed of sound). Even with a short barrel revolver of less than three inches, the .44 caliber will consistently hit a target the size of a man at 100 yards with tighter groups than with the .22 caliber bullet. Of course, a good marksman can do the same thing with a .22 caliber handgun with the only difference being larger groups at that same distance.

What happens after the bullet drops below the speed of sound depends on a huge variety of complex variables, everything from the marksmanship of the shooter to the amount of moisture in each increment of air density of the bullet's flight path.

We have a problem of definitively defining the maximum effective range. We know the maximum effective range begins when the bullet's trajectory velocity drops below the speed of sound. We also know, although it loses its coherence immediately, it can travel a considerable distance before we will have realized we can no longer hold onto a group. The most we can do, without extensive field experiments with each bullet, is to use the LTTE method (see 15.4), with the subexpressions of the drag nose compression and the drag cylinder compression, for surprising accuracy, to calculate the distance downrange when the bullet's predicted trajectory velocity drops below the speed of sound.

Experience has taught us a bullet can travel an enormous distance with serious lethality well beyond this threshold of prediction and still maintain a measure of coherence in its grouping individuality. This pattern of dispersion is always the same with each bullet, regardless of its weight or design. As it continues its movement downrange, after dropping below the speed of sound, the coherence of its groups gets bigger and bigger until the groups get too big to hit anything or it runs out of sufficient trajectory velocity to maintain a forward course. Under normal ballistic constraints, this means that any bullet of substantial weight and muzzle velocity can easily travel for miles until it either hits something or falls to the ground.

Below is Algorithm 3, a derivative of Algorithm 2. It uses the LTTE method to calculate the loss of velocity and kinetic energy in the bullet's flight path through a calculation of the loss of kinetic energy through transfer of energy, the first component of drag. Every time the bullet transfers another increment of its remaining kinetic energy in an increment of air in the bullet's flight path, it loses an increment of its forward trajectory velocity. Then, in this particular algorithm, it includes the subexpressions of drag through nose compression and the drag through cylinder compression to provide greater accuracy, usually well above 90%.

In the instructions on lines 370 and 380, the program continues to calculate the loss of velocity until the remaining velocity drops below the speed of sound (1130 fps) (obviously an arbitrary figure because we will almost never know the exact speed of sound in most instances) and then it stops to calculate the distance in yards to represent the maximum effective range.

ALGORITHM 3

```
100 Print : "Maximum Effective Range"
110 Print : Input "Muzzle Velocity (fps) = ______," Y
120 Print : Input "Bullet Weight (grains) = ______," W
130 Print : Input "Kinetic energy at the muzzle (ft/lbs) = ______," T
140 Print : Input "Succeeding increment of kinetic energy (ft/lbs) =
    _____," E
150 Count = 0
160 O = E * P
170 K = T – O
180 T = K
190 O = E * P : Y = SQR ((2*K) / W/225218)) : IF W < 100 THEN
    GOTO 260
200 A = 196436.542 ^ (1/2)
210 S = (A/Y^ (1/2) /.0000509 ^ (1/2))) ^2 'NSA DRAG
220 P = SQR ((S^2)* ((SQR ((2*K) / (W/225218))) ^2)) /1964636.542
    'DRAG COMPRESSION
230 H = 19646365.42 ^ (1/2)
240 J = (H/Y ^ (1/2) /.0000509 ^1/2))) ^2 'CSA DRAG COMPRES-
    SION
250 Z = SQR ((J^2) * ((SQR ((2 * K) / (W/225218))) ^ 2)) / 19646365.42
    : GOTO 320
260 A = (1964636.542) ^ (1/2)
270 S = (A/Y ^ (1/2) / .0000509 ^ (1/2))) * 3 'NSA'DRAG
280 P = ((S ^ 2 * Y ^ 2) ^ (1/3)) / 1964636.542
290 H = 19646365.42 ^ (1/2)
300 J = (H/(Y^ (1/2) / .0000509 ^ (1/2))) * 3
310 Z = ((J ^2 * Y ^ 2) ^ (1/3)) / 19646365.42 'DRAG COMPRESSION
320 K = T – O
330 V = SQR ((2 * K) / (W/225218))
340 PRINT "Remaining Kinetic Energy = _________(ft/lbs)
350 PRINT K
360 PRINT "Maximum Effective Range"
370 IF V > 1130 THEN COUNT = COUNT + 1 : GOTO 160 ELSE 380
380 IF V = 1130 THEN GOTO 390
```

```
390 DEFINT I : I = COUNT / 3600 * 100
400 PRINT "Maximum Effective Range = ______ yards"
410 PRINT I : PRINT CHR$(7)
420 END
```

Note: *To obtain "succeeding increment of kinetic energy," multiply the kinetic energy at the muzzle by .0000509 (dry air density) [TE = KE x σ] to get the TE and then subtract that answer from the KE at the muzzle. From there, you will get the remaining KE in the succeeding increment of penetration of the bullet's flight path.*

CHAPTER 22

THE CORRELIS EFFECT

"The apparent deflection (Coriolis acceleration) of a body in motion with respect to the earth, as seen by an observer on the earth, attributed to a fictitious force (Coriolis force) but actually caused by the rotation of the earth and appearing as a deflection to the right in the Northern Hemisphere and a deflection to the left in the Southern Hemisphere."

—Dictionary.com

"In physics, the Coriolis effect is a deflection of moving objects when the motion is described relative to a rotating reference frame. In a reference frame with clockwise rotation, the deflection is to the left of the motion of the object; in one with counter-clockwise rotation, the deflection is to the right. Although recognized previously by others, the mathematical expression for the Coriolis force appeared in an 1835 paper by French scientist Gaspard-Gustave Coriolis, in connection with the theory of water wheels. Early in the 20th century, the term Coriolis force began to be used in connection with meteorology."

—Wikipedia, the free encyclopedia

In the science of small arms ballistics, the Correlis Effect, except for military snipers and the serious long-distance competitive target shooters, is of no importance to us. It is an important phenomenon to study when we desire or need a more solid grasp of the science of ballistics, however.

"*Gustave-Gaspard Coriolis,* (born May 21, 1792, Paris – died Sept. 19, 1843, Paris), French engineer and mathematician who first described the *Coriolis force, an effect of motion on a rotating body, of paramount importance to meteorology, ballistics, and oceanography. An assistant professor of analysis and mechanics at the École Polytechnique, Paris (1816–38), he introduced the terms work and kinetic energy in their modern scientific meanings in his first major book, Du calcul de l'effet des machines (1829; 'On the Calculation of Mechanical Action'), in which he attempted to adapt theoretical principles to applied mechanics.'*"

—Written by: The Editors of Encyclopedia Britannica

Basically, in respect to the science of small arms ballistics, the Correlis Effect is the phenomenon of the Earth's rotation causing bullets to deflect to the right of its aiming point in the Northern Hemisphere and to the left in the Southern Hemisphere. In field artillery, coastal artillery, and ship artillery, this phenomenon becomes extremely important if we want to hit anything at great distances. A military combat ship, if it expects to hit another ship in combat moving in any direction relative to itself, must include the effect of the Correlis Effect in its calculations as well as the calculations the enemy ship will have travelled by the time the projectile reaches it.

In the army, shooting field artillery at an enemy emplacement at great distances of several miles to more than 20 miles, although these enemy emplacements may not move or move as fast as an enemy ship in sea, it is still necessary to include in its calculations the effect of the Correlis Effect in order to hit it. Otherwise, depending on the distance between the artillery and the enemy emplacement, the projectiles could miss by miles.

In small arm ballistics, the Correlis Effect becomes important when someone shoots a gun straight up in the air, but negligible at less than 1000 yards straight ahead. The bullets will lose forward velocity very quickly, and it is very unlikely such bullets in small arm ballistics will travel more than a few hundred yards up in the air, at the very most, before it begins to fall back to Earth.

Because of the Correlis Effect, however, in the Northern Hemisphere, the bullets will deflect to the right as it falls and may take a considerable amount of time tumbling in flight (depending on its weight) before it reaches ground several miles away from it origin and nowhere near its aiming point.

This ignorance of the Correlis Effect in small arm ballistics can lead to some interesting criminal endeavors. One particular and well-documented account occurred about 20 years ago when a greedy real-estate property developer, anxious to obtain ownership of approximately 170 acres of underdeveloped land in use by a shooting and fishing club, tried to destroy this club while thinking of himself as clever.

This club uses a small mountain, of at least 600 feet in height above the height of the shooting range, as its backdrop to stop bullets from causing harm to property and property owners on the other side of the mountain. With a high-power rifle and several rounds of ammunition, he climbed on top of the mountain to shoot down at houses and automobiles on the opposite side.

He held the purpose of wanting to make it look as if people on the other side shooting their guns in the shooting range were shooting a little too high, subsequently causing the bullets to strike personal property on the other side and with the potential of hitting children.

If true, sure enough, this would make the shooting range in the community very dangerous. Just think of the damage to personal property and the anguish to the parents of young children if one should get shot.

However, this greedy real-estate developer made one crucial mistake. He did not understand the physical phenomenon of the Correlis Effect in small arm ballistics. If anyone were shooting a little too high at this range, due to the Correlis Effect, the bullet could not possibly arrive on the opposite side of the mountain directly in front of the shooter.

Remember General Hatcher's experiments between the two world wars? The .30 caliber bullets would go up several hundred feet to several hundred yards in the air, depending on the bullets' weight and trajectory velocity, lose it forward velocity very quickly, and then slowly fall to ground while deflecting to the right of the aiming point, possibly landing several miles from the shooting position.

He then made a second crucial mistake. He made the mistake of talking about people reporting seeing bullet holes in their cars and houses, when not reported to the news media, and then expressing a burning desire to purchase the land himself as soon as it becomes available for sale.

The land was never sold, and the shooting range is still very much active today.

CHAPTER 23

TRUE MINUTE OF ANGLE

"The arcminute is commonly found in the firearms industry and literature, particularly concerning the accuracy of rifles, though the industry refers to it as minute of angle. It is especially popular with shooters familiar with the Imperial measurement system because 1 MOA subtends approximately one inch at 100 yards, a traditional distance on target ranges. Since most modern rifle scopes are adjustable in half (½), quarter (¼), or eighth (⅛) MOA increments, also known as clicks, this makes zeroing and adjustments much easier. For example, if the point of impact is 3" high and 1.5" left of the point of aim at 100 yards, the scope needs to be adjusted 3 MOA down, and 1.5 MOA right. Such adjustments are trivial when the scope's adjustment dials have an MOA scale printed on them, and even figuring the right number of clicks is relatively easy on scopes that click in fractions of MOA."

—Wikipedia, the free encyclopedia

The minute of angle is the unit of measurement to measure a gun's accuracy, usually rifles. Typically, under perfect or controlled conditions, with the gun mounted on a bench-rest, without or with minimal wind, to remove any possibility of a shooter's error, we use one minute of angle as the benchmark of accuracy. That means, if the rifle can shoot a group of one inch at 100 yards, it is arbitrated as a precision rifle. Some firearm manufactures advertise some of their precision rifles shoot groups smaller than one minute of angle at 100 yards, however,

$$\text{MOA} = 60\left[\text{Tan}^{-1}\left(\frac{\text{Group}}{\text{Ds}}\right)\right] \quad (1)$$

where Group = size of group (inches); *Ds* = distance between gun and target (inches).

SECTION THREE

CHAPTER 24

THE SCIENCE OF TERMINAL BALLISTICS

CONTENTS

"Terminal ballistics, (also known as wound ballistics) a sub-field of ballistics, is the study of the behavior and effects of a projectile when it hits its target and transfers its energy to the target. Bullet design and the velocity of impact determine the effectiveness of its impact.

"The study of terminal ballistics (also known as wound ballistics) is important to hunters to ensure that the animals they shoot are killed in as humane a fashion as possible. Ethical hunters strive to inflict a quick kill subjecting the animal to as little pain and suffering as practical. The firearm and cartridge are tailored to specific game animals toward this end.

In military and police use the study of terminal ballistics is often the study of how the impact of bullets affects human beings. The goal of the Policeman when he is compelled to fire on another person is to stop an imminent threat as soon as possible (not necessarily to kill). The goal of the soldier is similar to that of the policeman but the "rules of engagement" of these two occupations are considerably different. A soldier might be attacked from a long distance where the Policeman is often within a few yards of a threat.

"The study of terminal ballistics has as its goal the development of ethical solutions to the problems of firearms use against living targets."

—Wikipedia, the free encyclopedia

24.1 DEFINITION

Terminal ballistics is the scientific study of the patterns and relationships or interactions between the bullet and the target.

In the terminal ballistics of the bullet, we study the effect of the target over the bullet or the patterns and relationships of the transfer of energy from the bullet to the target.

In the terminal ballistics of the target, we study the effect of the bullet over the target or the patterns and relationships of the acceptance of penetration or denial (reflection) of energy from the bullet.

A study of terminal ballistics starts at the precise moment the bullet begins to transfer energy into the target; the target begins to accept or reflect energy from the bullet or the bullet begins to penetrate the target.

Such a study would also include the failure of the bullet to transfer energy into the target, a refusal of the target to accept energy from the bullet, or a failure of the bullet to penetrate the target.

CHAPTER 25

TRANSFER OF ENERGY

CONTENTS

"In the physical sciences, an energy transfer or 'energy exchange' from one system to another is said to occur when an amount of energy crosses the boundary between them, thus increasing the energy content of one system while decreasing the energy content of the other system by the same amount. The transfer is characterized by the quantity of energy transferred, which can be specified in energy units such as the joule (J), in combination with the direction of the transfer, which can be specified as in (to) or out of (from) one system or the other. The transfer occurs in a process which changes the state of each system."

—Wikipedia, the free encyclopedia

25.1 INTRODUCTION

In Figure 25.1, we have two components of physics to form the relationship of transfer of energy (*TE*):

(1) Kinetic energy (*KE*)
(2) Target density (*Tp*)

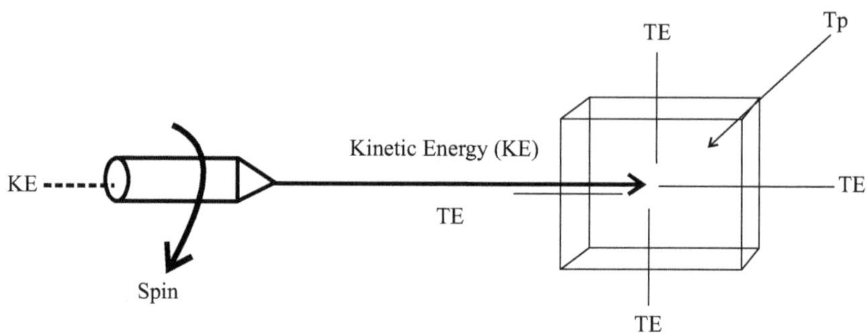

FIGURE 25.1

A careful study will easily reveal velocity, weight, mass, momentum, etc., as *not* the correct components or variables that make up the relationship of transfer of energy. It is only kinetic energy and target density.

- If we were to increase the bullet's kinetic energy, we will increase the transfer of energy into the target or anything else possessing density.
- If we were to increase the target's density, we will increase the transfer of energy from the bullet into the target.

Therefore, the relationship of transfer of energy is the *product* of kinetic energy (KE) and target density (Tp) or:

$$TE = KE \times Tp \tag{1}$$

As we can now see, the actual amount of kinetic energy a bullet can or will transfer into a target or anything with density depends directly on the amount of kinetic energy available in the bullet and the density of the target the bullet strikes. Without density, the bullet can only possess its *potential energy,* also a product of its mass and the square of its velocity, and working out the above relationship will readily prove the bullet cannot transfer all of its kinetic energy into the target, either. Some of it converts into heat, light, noise, and other electromagnetic phenomena, including radio waves. This means, among other things, each target density and bullet combination will provide its own distinctive and identifiable signature when the bullet strikes a target.

It also means, or at least suggests, that if we were to develop the science and technology, we could render either a bullet or an explosive charge inside of the bullet (projectile) harmless once we will have removed the kinetic energy from a passing bullet or the repercussive waves of an explosion (containing the lethal kinetic energy) in front of the target.

From the other end of this perspective, if we could manipulate target density, such as a reduction of volume for a given mass, we could just as easily either render it useless as a defense barrier with the approaching bullet or projectile with its explosive charge, or increase the density, through an increase in volume for a given mass, to render the target impenetrable.

A passing bullet or the repercussion waves of an explosion, either conventional or nuclear, will possess only a minimal capability to inflict damage if we were to remove its kinetic energy with some kind of force-field (evolving from the science of small arms ballistics). A target, or defense barrier, becomes transparent if – with the aid of a projecting force-field – we were to manipulate either a reduction or displacement of density through a manipulation of volume.

In Eq. (25.1), representing the physical relationship between transfer of energy to kinetic energy and density, it is a generalization that will only work in simple problems. If we were to attempt a calculation of transfer of energy into a given target, however, with the bullet still moving through it, it is not so simple. As we do with the calculation of the loss of kinetic energy and velocity, through the transfer of energy into each cubic inch of the bullet's flight path, using the LTTE method, we have to calculate the transfer of energy into the target at every cubic increment of penetration. Each subsequent penetration will transfer a little less energy into the target, precisely corresponding to a reduction of bullet terminal velocity and terminal kinetic energy.

A derivative of the LTTE method takes the following appearance:

$$TE\xi = \text{Lim} \int \sum_{\xi \to Ŧ}^{n=\infty} \left[p\Delta TE\xi - \left(\Delta KE\xi \times \Delta T\mathrm{p}\xi \right) \right] \tag{2}$$

where $TE\xi$ = transfer of energy in the initial terminal velocity (ξ); $\xi \to$Ŧ = computations from the initial terminal velocity (ξ) to the point the bullet stops entirely in the target (Ŧ) or leaves it.

Routine: *Similar to LTTE, take the second increment (Δ) of transfer of energy (TE) of the initial terminal velocity (ξ) and multiply it by the second* *increment (Δ) of kinetic energy (KE) in the initial terminal (ξ) and, with that answer, subtract it from the first or proceeding increment ($p\Delta$) of the initial terminal velocity. Continue to compute to each increment until the bullet stops in the target (Ŧ); then add (Σ) up each answer in each increment. It is not really complicated and rarely takes more than 6 to 12 iterations for living tissue.*

A manipulation of the transfer of energy relationship will allow us to calculate the kinetic energy at the initial terminal velocity (ξ) if we know the transfer of energy (*TE*) and the target's density (*Tp*):

$$KE = \frac{TE}{Tp} \tag{3}$$

Likewise, if we know the transfer of energy and the kinetic energy in the first increment of penetration, we can then calculate the target's density:

$$Tp = \frac{TE}{KE} \tag{4}$$

25.2 CONCLUSION

A study of the transfer of energy relationship in the science of terminal ballistics can be the beginning or inception of new sciences. Without a proper and solid grasp of this important subject, we could never readily understand terminal ballistics as a most significant science, or its ramifications, implications, and offshoots toward the development of other sciences and technologies.

Such knowledge can lead us to most significant developments, such as force-fields and ballistic signatures.

CHAPTER 26

TEMPERATURE OF TRANSFER OF ENERGY

CONTENTS

26.1 INTRODUCTION

As stated earlier, in the chapter on Transfer of Energy, part of the kinetic energy that transfers into the target, or anything with density, such as the air in the flight path, converts into heat as well as other electromagnetic phenomena.

- If we increase the bullet's velocity into a material consisting of any measure of density, we increase the temperature (T°) of the transfer of energy into that material.
- If we increase the density of the material to any given bullet velocity, we also increase the temperature of the transfer of energy into that material.

Therefore, the mathematical relationship of the temperature (*T*°) of the transfer of energy (*TE*) is the *product* of velocity (*V*) and density (*p*); or:

$$T^\circ = V \times p \quad (1)$$

Because this relationship of temperature of the transfer of energy is a generalization of the relationship between velocity and density, the above equation would only apply to the first increment of material density the bullet strikes. To calculate the temperature of the transfer of energy in the entire bullet's terminal flight path ($T°TE\xi$), we would need a special set of instructions where:

$$T^{\circ}TE = \text{Lim} \sum_{\xi \to \mp}^{n=\infty} \left(\Delta V \times \Delta Tp \right) \quad (2)$$

Each successive increment of density in the bullet's terminal flight path would possess a small but calculable lower temperature of the transfer of energy, due to the bullet slowing down, than in its preceding increment of density.

In many circumstances, it may be more appropriate, perhaps more useful statistically, to identify and measure, or calculate, the exact temperature of the transfer of energy at each increment of the terminal flight path and then to calculate either the *mean average* or the *median* (see Statistics).

Mean average is the sum of the values of each increment in the bullet's terminal flight path divided by its number of increments.

Median is the value in a distribution of values, of increments in the terminal flight path, representing a *mid-range* (50%) of the distribution of values.

Both mean average and median have their advantages and, of course, disadvantages. Whether we use mean average or median depends on the actual circumstances, the nature of the material density and densities, depth of density or densities, and the distribution pattern of values in the terminal flight path.

26.2 MANIPULATION OF THE RELATIONSHIP

If we know the temperature of the transfer of energy and the bullet's terminal velocity, we can calculate the density at each increment of penetration:

$$Tp = \frac{T^{\circ}}{V} \quad (3)$$

On the other hand, if we know the temperature of the transfer of energy at any given increment and the material density at that increment, we can calculate the bullet's velocity:

$$V = \frac{T^{\circ}}{Tp} \tag{4}$$

26.3 CONCLUSION

We can use this important relationship as a tool to work out more complex problems in terminal ballistics and as part of a "building block" to our knowledge on the entire subject.

By itself, knowledge of the temperature of the transfer of energy may have little use to us presently; however, some time in the future, it may become critically important in certain military applications. In one application in mind, if we know the density of a given target and the temperature of the transfer of energy in the first increment of penetration, we could then calculate the initial terminal velocity and, with that knowledge, we could proceed to calculate the bullet's trajectory *backwards* to determine the point of origin, one of the many techniques to locate and identify a sniper's position.

CHAPTER 27

REFLECTION OF KINETIC ENERGY

CONTENTS

27.1 INTRODUCTION

"*Reflection is the change in direction of a wavefront at an interface between two different media so that the wavefront returns into the medium from which it originated. Common examples include the reflection of light, sound and water waves. The law of reflection says that for specular reflection the angle at which the wave is incident on the surface equals the angle at which it is reflected. Mirrors exhibit specular reflection.*"

—Wikipedia, the free encyclopedia

Reflection of kinetic energy (*RE*) represents the relationship between the kinetic energy (*KE*) at the initial terminal velocity (ξ) to the kinetic energy not remaining in the bullet or travelling in the target. It is the energy that reflects from the target and no longer accompanies the bullet and consist of the electromagnetic phenomena we associate with heat, light, noise, radio waves, etc., where:

$$RE = KE\xi - KE \tag{1}$$

where RE = Reflected energy (ft/lbs); $KE\xi$ = Kinetic energy at the initial terminal velocity (ft/lbs); KE = Kinetic energy not remaining in bullet or in target (ft/lb)

A simple manipulation of the above algebraic equation will help us to calculate the kinetic energy at the initial terminal velocity if we know the reflection of energy and the kinetic energy not remaining in the bullet or in the target:

$$KE\xi = RE + KE \qquad (2)$$

With the above knowledge, we can proceed to calculate the initial terminal velocity:

$$\xi = \sqrt{2KE / M} \qquad (3)$$

Then, with knowledge of the weapon and type of ammunition, we can proceed to calculate the bullet's trajectory *backwards* in order to identify and locate the shooter's origin.

27.2 CONCLUSION

Reflection of energy is another "building block" in our knowledge of terminal ballistics. Though it does not necessarily seem terribly important by itself, it can lead us to analyze and work out solutions to problems in terminal ballistics and certainly to provide each of us with a stronger line of continuity of insight and comprehension of this whole subject of the science of small arms ballistics.

Reflection of energy is the measurable reflective signature each bullet creates as it strikes anything with density.

Anything with density will leave a distinctive and easily identifiable signature we can use to identify the target's composition, density, and temperature or the bullet's composition, density, temperature, velocity, or kinetic energy, including the electromagnetic phenomena we associate with heat, light, noise, radio waves, etc.

CHAPTER 28

ACCEPTANCE OF KINETIC ENERGY

CONTENTS

28.1 INTRODUCTION

In the study of the acceptance of kinetic energy, our primary concern ought to be the target's ability to accept energy from the bullet. Though any bullet can easily transfer energy into anything or any target with density, the target itself cannot accept the bullet's energy unless the bullet at least partially penetrates it. Then, with penetration, the amount of energy the target accepts (*AE*) is the summation of the transfer of energy into each increment of penetration in its terminal flight path:

$$AE = \sum_{\frac{3}{4} \rightarrow \mp}^{n=\infty} \Delta TE \tag{1}$$

Because each subsequent increment of penetration would possess a different and lower value of kinetic energy the bullet can transfer into the target, as it penetrates the target, we have the same relationship as we do with the transfer of energy.

28.2 CONCLUSION

Acceptance of energy deals with the study of the target's ability to accept energy from the bullet as it penetrates the target, although it may or may not cause damage to the target.

Without penetration, there can be no acceptance of energy from the bullet to the target or damage to the target for that matter.

CHAPTER 29

THEORY OF PENETRATION

CONTENTS

29.1 INTRODUCTION

If we were to fire the Colt M-16, the Soviet-made AK-47, and Smith & Wesson's M-29 at a metallic target of any particular density but of sufficient density to eventually stop the bullet, a pattern will eventually emerge to demonstrate a relationship between velocity and bullet diameter to bullet density and target density – to penetration.

The M-16 fires a 55 grain bullet of 0.224 inches in diameter at approximately 3,250 fps; the AK-47 a 122 grain bullet of 0.308 inches in diameter at nearly 2,250 fps, and the M-29 a bullet of 240 grains of 0.429 inches in diameter at a maximum muzzle velocity of 1,600 fps (with the right load and gun). Yet, each gun produces an average level of kinetic energy of between 1,300 and 1,400 ft/lbs at the muzzle (under optimum conditions). Though of different diameters, each bullet is made up of the same materials and densities.

29.2 THEORY

From this knowledge in the above Introduction, we have enough information to perceive a pattern and relationship between velocity, diameter, and density, where penetration (*P*) represents:

$$P = \frac{Bp(V/d)}{Tp} \tag{1}$$

where Bp = bullet density (lbs/cubic inch); V = velocity (fps); d = diameter of bullet (inches); Tp = target density (lbs/cubic inch).

- When we increase bullet density, we *increase* its penetration.
- When we increase bullet velocity, we *increase* penetration.
- When we increase bullet diameter, however, we *decrease* penetration.
- When we increase target density, however, we *decrease* bullet penetration.

From these patterns and relationships, we can easily derive additional conclusions through a manipulation of the above algebraic equation.

If we take the bullet density (*Bp*) and place it to the left of the equal (=) and penetration (*P*) to the right, we can determine the best bullet density for a given target density, bullet velocity, and diameter:

$$Bp = \frac{Tp}{\left[(V/d)P\right]} \tag{2}$$

Now it becomes obvious, if we can manipulate the equation for penetration to determine the best bullet density for a given set of circumstances, we can just easily manipulate the same equation to determine the best bullet velocity for a given diameter, bullet density, and target density:

$$Bv = \frac{Tp}{\left[(Bp / d) / P\right]} \tag{3}$$

Then, if we know the bullet density, velocity, diameter, and depth of penetration into a given target, we can easily calculate the target's density:

$$Tp = \frac{\left[Bp(V / d)\right]}{P} \tag{4}$$

At the same time, if we must design-engineer a gun-system for a particular target density with certain known bullet density and velocity parameters, we can now easily calculate the best bullet diameter (Bd) to successfully penetrate the target, either partially or completely:

$$Bd = \frac{\left[Bp(V / Tp)\right]}{P} \tag{5}$$

29.3 APPLICATION

These relationships dealing with a study of penetration apply to hunting as well. However, instead of changing the metal or lead alloy to control and vary density for a given hunting application, the firearms accessory industry manipulates the bullet's density through a manipulation of its jacket thickness and composition. If greater density becomes necessary to penetrate something as dense as the armor-plating material on a tank, then the bullet engineers often will simply switch to a denser core material such as depleted uranium, and – if realistically possible – either increase the muzzle velocity or drop the bullet's diameter. Usually, the engineers lack the option to change diameter; they must ordinarily work with a particular gun-system which may already operate at its maximum potential in its present caliber. So, an increase in the bullet's composition becomes the only option available to them other than to work out a new gunpowder(s) formula to produce a greater muzzle velocity with the same chamber pressure, if possible.

In the sporting firearms accessory industry, when targets tend to be much less dense than a tank's armor, the bullet engineers have even more technical restrictions as they seek out a solution to each problem. Except for special, non-hunting applications such as indoor target practice, a lead core is absolutely mandatory. It is not only the recent laws from the United States Congress, local or state governments, or the people who make a career out of the gun control controversy when not competent to discuss the subject, or the people who latch onto an ideology as a substitute for intelligence, that complicates the restrictions toward a better solution, but simply a problem in the availability and suitability of the selection of materials and their physical properties. We only have so many materials suitable for a bullet and, frankly, lead or a lead alloy (consisting of tin and alimony principally) makes the best bullets ordinarily.

29.4 USE OF BULLET LUBRICANTS

However, as we increase the muzzle velocity, from a few hundred feet per second to several thousand feet per second, we must lubricate the bullets to prevent them from leading the bore, whether by using bear grease, an elaborate formula of beewax and alox 2137F, a glassine or grease-proof paper or an alloy of copper-zinc around a lead core. When we increase muzzle velocity, we must improve the lubricant in order to avoid leading the bore each time. As lead builds up in the bore, with each subsequent shot, it incrementally increases the resistance to the bullet's movement in the bore, with a corresponding increase in bore pressure, a reduction in bore velocity and, of course, it changes the time-pressure curve in each subsequent shot to knock it out of time.

With each subsequent shot, and with each successive build-up of lead in the bore, the gunpowder behind the bullet, as it burns traveling through the bore, will stop burning further and further away from the muzzle. At the same time, we will notice the groups getting progressively bigger downrange until, eventually, maintaining a group or hitting anything consistently becomes a real challenge.

Bear grease is predominately a low-velocity lubricant, used in black powder muzzle-loaders, and copper-zinc jackets serve as a high-speed lubricant for high-power center-fire rifles and handguns. When not using enough lubricant, if we were to look carefully, we will see a change in the color of the smoke coming out of the bore, probably a combination of blue/gray/black out to 5, 10, or 15 feet or more.

Likewise, when using too much lubricant, we will see a similar change in the color of the smoke plus a much stronger and a most distinctive, unforgettable odor of lubricant. All this is particularly profound when shooting in the snow. Whether too much or too little lubricant, it leaves a most distinctive trail in front of the bore.

29.5 MANIPULATION OF DENSITY TO INCREASE VELOCITY AND PENETRATION

In time, though first and foremost a trial-and-error system, a variety of people, inside and outside of the firearms industry, among professionals and amateurs, discovered that if they were to place the lead core inside of a copper-zinc jacket and then to vary the thickness of the jacket for a given target, they could vary its penetration and hence effectiveness into that given target, whether for squirrels or a dangerous rhinoceros.

More likely than not, they did not realize they were simply manipulating the bullet's density through a manipulation of the jacket's thickness. This manipulation of density would control its penetration. Density, as we discovered in the relationship described in the algebraic formula of penetration, is one of the three variables responsible for bullet penetration.

29.6 VARMINTING

In varminting – shooting at small game – it became necessary to develop guns and bullets that could shoot very tight groups at extremely long distances while maintaining velocities above the speed of sound. Though some small animals may be more intelligent than others, varminters soon discovered all of them were intelligent enough to perceive danger at distances of several hundred yards and to run and hide in their underground homes upon sight of a man with a gun, which made hunting them an exceedingly difficult and complex technological challenge, particularly in the open grasslands of Oklahoma and similar terrain throughout the world.

At first, these hunters, in full recognition of the technical challenge, demanded better guns and, later, better triggers assemblies, optical sights, sighting scopes, bullets, and range-finders.

It is beyond the scope of this text to describe the developmental problems of guns, barrels, triggers, optical range-finders, and sighting systems. So, to restrict ourselves to ballistics, we will stay with the study of the bullet.

Because these small critters were really very, very small, they were hard to see at great distances on the grasslands and even more so to hit with conventional arms. When the hunter finally did strike one of them, he discovered they did not die easily, either. To make it worse, they would typically survive long enough to run, crawl, creep, or edge back into their underground holes, either making it impossible to confirm a hit or to fetch the game to eat later.

Those bullets in use at the time provided far too much penetration to frequently cause massive damage to the animals without killing them immediately, giving them time to run and hide.

Eventually, some of these hunters recognized the problem as a problem with the design of the bullet. Their jackets were simply too thick to cause the bullets to completely penetrate the animals, sometimes with only superficial damage. When these bullets did penetrate completely through the animals, much of the lethal kinetic energy we need and use to kill it went with the bullet to strike something behind the animal, usually Motherearth. Cast bullets, though certainly cheap enough, also have the tendency to penetrate completely to cause very minimal and sometimes superficial tissue damage.

Using a trial-and-error system, bullet manufacturers, to meet the demands for better varminting bullets, began to produce bullets with very thin jackets. They discovered, at very high velocities – from 2500 to 4000 fps – the bullets would expand violently within an inch or two of penetration. This explosive collision would effectively prevent or slow down complete penetration and therefore transfer most of the remaining kinetic energy from the initial terminal velocity into the animal.

Though bullet engineers and manufacturers may have never used a mathematical system to determine the correct jacket thickness, it is nevertheless entirely possible to arrive at a very realistic conclusion with the following equation:

$$Vthj = \frac{SD^2}{M \times \xi} \qquad (6)$$

where *Vthj* = Jacket thickness for a varminting bullet (inches); *SD* = sectional density (look for it in reloading manuals); *M* = Bullet's mass (weight/225,218 = mass); and ξ = expectant terminal velocity at the target (fps).

"*Sectional density is the ratio of an object's mass to its cross-sectional area. It conveys how well an object's mass is distributed (by its shape) to overcome resistance. For illustration, a nail can penetrate a target medium with less force than a coin of the same mass. During World War II bunker-busting Röchling shells were developed by German engineer August Cönders, based on the theory of increasing sectional density to improve penetration. Röchling shells were tested in 1942 and 1943 against the Belgian Fort d'Aubin-Neufchâteau and saw very limited use during World War II. Sectional density is often used in gun ballistics where sectional density is the ratio of a projectile weight to its diameter*"

—Wikipedia, the free encyclopedia

$$SD = \frac{W}{7000 \times d2} \qquad (7)$$

where *SD* = sectional density[1]; *W* = weight of bullet (grains); *d* = diameter of bullet (inches).

---

[1] Note: *Beware! There is considerable fraud and controversy in the proper use and definition of sectional density. With the association of sectional density to greater range and accuracy, some bullet manufacturers exaggerate their sectional density figures to encourage the ammunition reloaders to use their bullets instead of their competitor's, although they claim in their literature they use the above formula to calculate sectional density. Not always true!*

29.7 BULLETS FOR BIG-GAME HUNTING

As the hunter went after bigger and bigger game, it became increasingly obvious for him, in order to increase penetration in the game, he would need to increase the jacket thickness. For a reduction in diameter was out of the question, in most instances. Either he had to stick to the same caliber (diameter) or was committed to it due to economic considerations. Changing the caliber can be very expensive.

From experience, if the bullet expands too quickly, due to the thinness of the jacket, it would waste the dispersion of kinetic energy on the surface of the animal to create very large but superficial wounds. Not enough of the kinetic energy would enter the interior of the animal to be sufficiently lethal. Due to the size of these big-game animals, an arbitrary figure of 200 ft/lbs per cubic inch of penetration for a minimum of 6 to 12 inches in the animal's torso has been chosen as a realistic figure in most instances, depending on skin depth and thickness and its physiological makeup at the time of the encounter. No different than man, animals are hard to kill!

If, on the other hand, the jacket was too thick, the bullet would penetrate completely to carry most of the remaining kinetic energy with it. Pursuant to the transfer of energy relationship, a significant but not necessarily an adequate amount of lethal kinetic energy would transfer into each increment of penetration; instead, most of this lethality would leave the animal on the other side of its torso, rendering the animal still alive either running away to live for another day or to die somewhere out in the field from the combination of trauma and the lost of blood. In some instances, the poor animal may run a few hundred yards just for another hunter to take credit for shooting him as well as the privilege to take home his meat.

Finally, both the hunter and manufacture have recognized a need to use a jacket with just the right thickness to allow the bullet to penetrate more deeply than a varminting bullet but not completely. They wanted the bullet to penetrate about half-way into the torso of the animal's body and then stop in order to transfer all the remaining kinetic energy for the maximum effect. The algebraic equation below will give the approximate jacket thickness of these hunting bullets:

$$HBthj = \frac{SD^2}{2} \tag{8}$$

where *HBthj* = Heavy Bullet jacket thickness (inches); *SD* = sectional density.

29.8 DANGEROUS GAMES

Later, while shooting big and really dangerous game, it became obvious these hunters needed a bullet with a greater jacket thickness and a much heavier weight to survive the physical shock of entering massive tissue depth without deforming beyond recognition and while penetrating much more deeply to transfer all or most of its remaining lethal kinetic energy. Below is an algebraic equation to provide for an approximate answer to this problem, of course depending on the distance the bullet must travel before it reaches the game and its initial terminal velocity, which should be above the speed of sound to be effective.

$$DGthj = \frac{d(SD)}{2} \tag{9}$$

where *DGthj* = Dangerous Game jacket thickness (inches); *SD* = sectional density; *d* = Bullet diameter (inches).

Now, we must realize it was not the original purpose of bullet engineers and manufactures to use the copper-zinc alloy jackets around the lead core for greater penetration, though it certainly did just that. They were using these jackets, perhaps in some instances without any apparent awareness of some side-effects, as a high-speed lubricant. At the velocities they were trying to obtain, they needed better lubricants to prevent lead from accumulating in the bore and, with these higher velocities; they would obtain a much flatter trajectory. With a flatter trajectory comes a lesser parabolic trajectory and hence a target easier to hit at varying distances with less compensation and sighting error.

It did not take long to recognize these higher velocities caused much greater penetration at longer distances as well as a flatter trajectory. Without understanding the relationship of penetration to bullet diameter, velocity, and density, they did not immediately grasp the significance of their altering bullet density through the use of a jacket around the lead core. When we re-examine the algebraic equation for penetration, the relationship between bullet velocity, diameter, and density becomes clearer.

29.9 A CLARIFICATION OF THESE RELATIONSHIPS

As some hunters began to hunt big and dangerous game with these high-velocity jacketed bullets, instead of the traditional paper patch bullets in those days, operating with substantially lower muzzle velocities, they discovered some undesirable side-effects: – Some bullets with their thin jackets would break up too quickly leaving the lead core separated from the deformed and fragmented jackets and others, with their thicker jackets – would simply completely penetrate the entire animal with little ill effect and minimal damage. If the bullet were to break up prematurely, the kinetic energy would scatter all over on the surface or in the shallow depth; if not, then the bullet may leave the animal entirely carrying most of the lethal kinetic energy with it. Hence, it was imperative for the bullet to stop and stay inside of the animal for the maximum transfer of energy into the torso to be truly effective, particularly for large and dangerous games. Else, we might find them going after us treating us as their game.

Then, some people, without a proper grasp of this subject, began to perceive, or thought they did, that if they were to design the bullet to expand as it travels through the body of the animal, it would transfer more or most of its lethal kinetic energy to cause the maximum kinetic upset. But this is not so! Transfer of energy from the bullet to the target is a function of the relationship between the bullet's kinetic energy and the target's density. Bullet diameter is not a variable in this relationship.

An expansion of the bullet as it travels through the animal or anything with density cannot increase the transfer of energy.

However, with diameter as a variable in the relationship of penetration, the expansion *increases* diameter and therefore *decreases* penetration as it slows down due to an *increase* in the bullet's nose and cylinder drag compressions with a parallel reduction in velocity, transfer of energy and nose and cylinder drag compression. If a successful design, the bullet will stop inside of the animal to transfer all its remaining kinetic energy for the maximum kinetic disruption and the transfer of energy inherent in that bullet design as long as the bullet stays intact. If it breaks up in terminal flight, it loses its effectiveness. It is not enough to simply design the bullet to expand upon impact with the target or within a certain distance after its first increment of penetration.

If the bullet, with its larger diameter due to expansion in terminal flight, fails to stay inside of the animal, it will not transfer any more kinetic energy than a non-expansion bullet penetrating completely; this is because it cannot transfer any more energy than it has available owing to the target's density. Remember! Transfer of energy is a function of the bullet's kinetic energy in flight (in motion) to the product of the target's density.

In fact, an expansion bullet may actually prove *less effective* as it slows down and finally completely penetrates. When we add up the total loss of kinetic energy through a transfer of energy, we may very well discover a non-expansion bullet may transfer more energy due to a higher average terminal velocity. Unless we can use the rapid increase in the bullet's diameter to eventually stop the bullet inside of the animal to prevent its complete penetration, contrary to popular belief, an expansion bullet will have no great advantage over a non-expansion bullet with the same level of kinetic energy at the same initial terminal velocity.

29.10 CONCLUSION

A study of penetration and its mathematical relationships reveal the ready possibility of manipulating penetration through a manipulation of:

- The bullet's density (through a manipulation of jacket thickness and composition).
- The bullet's diameter.
- The bullet's velocity.
- Once we know and understand the relationships of penetration, we can easily manipulate these relationships in order to calculate:
- Bullet density.
- Bullet diameter.
- Bullet velocity.
- Target density.

Though the firearms industry, as a whole, does not use a mathematical system to determine jacket thickness, a trial-and-error system has been in use for years and works well sometimes.

An expansion bullet, unless it stops inside of the animal's body, cannot transfer more energy than a non-expansion bullet. Hence, it is necessary to manipulate the combination of every variable of penetration, including the jacket's thickness and the bullet's composition, in order to keep the bullet inside of the animal to be effective.

Though not mentioned above, there is also a safety and civil rights issue. When we hunt, shoot in self-defense or shoot in combat, we would prefer the bullets not to travel any further than the intended target.

CHAPTER 30

CALCULATING THE EXPECTANT DEPTH OF PENETRATION INTO ANIMAL ISSUE

CONTENTS

30.1 INTRODUCTION

We have discovered, in our study of the Transfer of Energy, the Theory of Penetration and the subsequent theories and methods to calculate trajectory, and we have also developed an inherent method of calculating the expectant depth of penetration into a target, whether animate or inanimate.

But it is a little complex and requires careful thought and methodology in its application.

30.2 SOLUTION

To resolve this problem, we need to know the muzzle velocity and the bullet's weight to determine its kinetic energy at the muzzle, the distance between the gun and the target, the initial terminal velocity, and the density of tissue to determine its depth of penetration.

If the bullet has the proper physical attributes to completely penetrate the target, we can also proceed to calculate the maximum expectant range of lethality beyond the target.

It will not take much thought for us to realize we shall need a very high-speed computer to handle this kind of number-crunching problem, as we do in the complex calculations of a bullet's trajectory.

There is an additional problem with the actual figure representing tissue density. What is the average density of living tissue – say, per cubic inch?. It seems to be somewhere between 0.02 and 0.04 pounds per cubic inch and, due to its *compressibility* characteristics, tissue density increases with bullet penetration.[1] The two components of drag, due to the surface of the nose and drag due to the surface of the cylinder, will naturally compress the tissue density immediately in front of the nose, slowing down the bullet even faster.

First, we must deal with two issues here. Whether the tissue density is 0.02 lbs/cu./inch or 0.04 lbs/cu./inch, or somewhat elsewhere, does not matter. Because the number of iterations in our computation is so small, typically between 1 and 18 for 1 and 18 inches of penetration, respectively, the accumulative error is completely negligible.

Secondly, the issue of compressibility is a function of pressure in front of the bullet, as defined in the science of thermodynamics and fluid mechanics. However, because the bullet, as it continues its penetration into living tissue, rapidly decelerates, the rate of tissue compressibility and density also rapidly decelerates to correspond with the bullet's rapid deceleration.

It would match the mathematical mechanics of the LTTE method in which each successive computation would be successively smaller than the preceding computation except, under very special circumstances, would not add up to anything of significance. The only thing of interest, which may or may not have any value, would be the number-crunching computations of calculating the accumulative compressibility values of every cubic inch of air in the bullet's flight path, or 3,600 iterations of computation for every interval of 100 yards.

---

[1] *"In thermodynamics and fluid mechanics,* ***compressibility*** *is a measure of the relative volume change of a fluid or solid as a response to a pressure (or mean stress) change."*

—Wikipedia, the free encyclopedia

Below is Algorithm 4, the first of two algorithms in this chapter, to calculate the expectant depth of penetration into living tissue using the LTTE method, namely the repetitive calculation of the loss of velocity and kinetic energy, through the transfer of energy from the bullet to each increment of penetration in the terminal flight path and, then, to either determine the depth of penetration or, if it penetrates completely, the expectant maximum range of lethality beyond the target. As with the other algorithms in this text, it was written originally in 1990 using Basic as the language on the Tandy 2000.[2]

ALGORITHM 4[2]

```
30 CLS : CLEAR : PRINT "Expectant Depth of Penetration into Liv-
    ing Animal Tissue"
40 PRINT : INPUT "Kinetic Energy at the muzzle = _____," T
50 PRINT : INPUT "Succeeding Increment of Kinetic Energy =
    _____," E
60 PRINT : INPUT "Tissue Density (lbs/cu./in.) = _____," P
70 PRINT : INPUT "Maximum Depth of Penetration (inches) =
    _____," Inches
80 PRINT : INPUT "Bullet Weight (grains) = _____," W
90 FOR I = 1 TO INCHES
100 O = E * P
110 K=T-O : T=K
120 O=E*P : Y=SQR((2*K)/(W/225218))
130 K=T-O
140 CLS 'Clear the screen
150 LOCATE 8,18 : PRINT I,Y,K
160 NEXT I
```

[2] Note: *To obtain "succeeding increment of kinetic energy," multiply the kinetic energy at the muzzle by .0000509 (dry air density) [TE = KE x σ] to get the TE and then subtract that answer from the KE at the muzzle. From there you will get the remaining KE in the succeeding increment of penetration of the bullet's flight path.*

```
170 IF Y< = 1 THEN LOCATE 12,20 : PRINT "DO NOT EXPECT
    FULL PENETRATION"
180 IF Y >1 THEN LOCATE 12, 20 : PRINT "EXPECT FULL ANI-
    MAL PENETRATION"
190 GOTO 240
200 LOCAE 14,20 : PRINT "DEPTH OF PENETRATION = _____
    INCHES"
210 LOCATE 14,42 : PRINT I
220 PRINT CHR$(7) 'BEEP
230 END
240 COUNT = COUNT + 1
250 O = E * 0.0000509
260 K = T – O : T = K
270 O = E * 0.0000509 : Y = SQR((2*K)/(W/225218))
280 K = T – O
290 I = I + 1
300 LOCATE 20,18 : PRINT COUNT, Y, K
310 IF K => 60 THEN COUNT = COUNT + 1 : GOTO 240 ELSE 320
320 IF K < 60 THEN GOTO 330
330 I = COUNT/3600*100
340 LOCATE 16,8 : PRINT "DISTANCE OF MOVEMENT BEYOND
    TARGET = "
350 LOCATE 16,60 : PRINT "YARDS"
370 PRINT CHR$(7) 'BEEP
380 LOCATE 24, 1, 1
390 END
```

The above algorithm calculates the depth of penetration into living tissue, using the LTTE method, starting at the initial terminal velocity. If these calculations determine the bullet will completely penetrate the target, then it continues on to calculate the expectant distance of lethality beyond the target, stopping when the calculations drop below 60 ft/lb of kinetic energy, the threshold of lethality for most humans under normal average conditions. It only calculates the transfer of energy drag, not the other two components of drag.

Algorithm 5 includes the calculations of all three components of drag: transfer of energy, nose drag compression, and cylinder drag compression.

It provides a greater accuracy for the calculations beyond the target; however, due to the small number of iterations, the difference between the two algorithms in the calculation of penetration is negligible. Neither algorithm attempts to handle the calculations or predictions of compressibility as the bullet moves through the living tissue.

ALGORITHM 5[4]

```
30 CLS : CLEAR : PRINT "PREDICTION OF LIVING TISSUE PENETRATION"
40 PRINT : INPUT "Kinetic Energy at the muzzle = _____," T
50 PRINT : INPUT "Succeeding Increment of Kinetic Energy = _____," E
60 PRINT : INPUT "Bullet's Weight (grains) = _____," W
70 PRINT : INPUT "Distance in Inches = " , INCHES
80 PRINT : INPUT "Maximum Depth of Penetration into Game (inches) = _____," Depth
90 PRINT : INPUT "Living Tissue Density (lbs/cu./in.) = _____," DENSE
100 DEFDBL P, S
110 CLS : LOCATE 12, 25 : PRINT "PLEASE STAND BY .... COMPUTING!"
120 FOR I = 1 TO INCHES
130 O=E*P
140 K=T-O : T=K
150 O=E*P : Y = SQR((2*K)/(W/225218)) : IF W < 100 THEN GOTO 190
160 A = 1964636.542^ (1/2)
170 S = (A/(Y^ (1/2)/.0000509^ (1/2)))^2
180 P = SQR((S^2)*((SQR((2*K)/(W/225218)))^2))/1964636.542 : GOTO 220
```

[4]Note: *When dealing with large, potentially dangerous game, in Algorithm 4, change line 310 to read, "IF K=> 200 THEN COUNT = COUNT + 1 : GOTO 240 ELSE 320" and line 320 to read, "IF K< 200 THEN GOTO 330." In Algorithm 5, change line 490 to read, "IF K=> 200 THEN COUNT = COUNT + 1 : GOTO 420 ELSE 500" and line 500 to read, "IF K < 200 THEN GOTO 510."*

```
190 A = (1964636.542)^ (1/2)
200 S = (A/(Y^ (1/2)/.0000509^ (1/2)))*3
210 P = ((S^2*Y^2)^ (1/3))/1964636.542
220 K=T-O
230 LOCATE 14,37 : PRINT I
240 NEXT I
250 PRINT CHR$(7) 'BEEP
260 FOR J = 1 TO DEPTH
270 O = E * DENSE
280 K = T – O : T = K
290 O = E * DENSE : Y = SQR((2*K)/(W/225218))
300 K = T – O
310 CLS 'CLEAR THE SCREEN'
320 LOCATE 8,18 : PRINT J, Y, K
330 IF K = < 1 THEN GOTO 350
340 NEXT J
350 LOCATE 12,16 : PRINT "*** DO NOT EXPECT FULL PEN-
    ETRATION ***"
360 IF K > 1 THEN LOCATE 12,16 : PRINT "*** EXPECT FULL
    PENETRATION ***"
370 LOCATE 14,10 : PRINT "DEPTH OF PENETRATION = INCHES
    OUT OF INCHES"
380 J = J – 1
390 LOCATE 14,32 : PRINT J
400 LOCATE 14,51 : PRINT DEPTH
410 PRINT CHR$(7) : IF K = 1 THEN GOTO 550
420 COUNT = COUNT + 1
430 O = E * .0000509
440 K = T – O : T = K
450 O = E * .0000509 : Y = SQR((2*K)/(W/225218))
460 K = T – O
470 I = I + 1
480 LOCATE 20,18 : PRINT COUNT, Y, K
490 IF K => 60 THEN COUNT = COUNT + 1 : GOTO 420 ELSE 500
500 IF K < 60 THEN GOTO 510
510 I = COUNT/3600*100
```

```
520 LOCAE 16,4 : PRINT "DISTANCE OF LETHALITY AFTER
    FULL PENETRATION = _____"
530 LOCATE 16,68 : PRINT "YARDS"
540 LOCATE 16,57 : PRINT I
550 PRINT CHR$(7) 'BEEP FOR COMPLETION OF PROGRAM
560 LOCATE 24, 1, 1
570 END
```

CHAPTER 31

BALLISTIC REFLECTION COEFFICIENT*

"In physics and electrical engineering the reflection coefficient is a parameter that describes how much of an electromagnetic wave is reflected by an impedance discontinuity in the transmission medium. It is equal to the ratio of the amplitude of the reflected wave to the incident wave, with each expressed as phasors."

—Wikipedia, the free encyclopedia

Ballistic reflection coefficient (BRC) deals with the proportion or ratio of the difference between kinetic energy at the initial terminal velocity ($KE\xi$) and the reflection of kinetic energy (RE) from the surface of the target to the sum of the kinetic energy at the initial terminal velocity and the reflection of kinetic energy from the surface of the target, where:

$$BRC = \frac{KE\xi - RE}{KE\xi + RE} \tag{1}$$

* Suggestion: *See a thesis on a report #1775, written by Nathan Gerber in August of 1966, with the Ballistic Research Laboratories dealing with a study of Ballistic Reflection Coefficient. Frankly, it is difficult to follow. You will find it via an Internet search engine. Please type "Ballistic Reflection Coefficient" in the search window.*

CHAPTER 32

BALLISTIC REFLECTION POWER

Ballistic reflection power (BRP) deals with the measurement of the target's power or ability to reflect kinetic energy and thereby reject penetration from a given bullet at a given initial terminal velocity (ξ), and therefore represents the product of reflection of energy (*RE*) and the differential of the ballistic spectrum ($d\beta s$), where

$$BRP = RE \times d\beta s \quad (1)$$

Differential or the degree of difference of the ballistic spectrum is the area of coverage between the .17 caliber and the big 16 inch guns on the USS New Jersey in which a target will successfully defend. It is the measurement of the power to reflect power, such as a bullet, without it causing harm.[1]

When we increase the *RE*, we increase our ballistic reflecting power. When we increase our differential ballistic spectrum or ability to deflect an attempt at penetration, we increase our BRP.

---

[1] *"The user[s] reflect any power that is launched against them without being harmed. This serves as a means of self-defense and counter-attack. The user can reflect whatever power is projected by their target and turn it back on them."*

—Superpower Wiki

CHAPTER 33

BALLISTIC PENETRATING POWER

"For use against armored targets, or large, tough game animals, penetration is the most important consideration. Focusing the largest amount of momentum on the smallest possible area of the target provides the greatest penetration. Bullets for maximum penetration are designed to resist deformation on impact, and usually are made of lead that is covered in a copper, brass, or mild steel jacket (some are even solid copper or bronze alloy). The jacket completely covers the front of the bullet, although often the rear is left with exposed lead (this is a manufacturing consideration: the jacket is formed first, and the lead is swaged in from the rear).

"For penetrating substances significantly harder than jacketed lead, the lead core is supplemented with or replaced with a harder material, such as hardened steel. Military armor-piercing small arms ammunition is made from a copper-jacketed steel core; the steel resists deformation better than the usual soft lead core leading to greater penetration. The current NATO 5.56mm SS109 (M855) bullet uses a steel-tipped lead core to improve penetration, the steel tip providing resistance to deformation for armor piercing, and the heavier lead core (25% heavier than the previous bullet, the M193) providing increased sectional density for better penetration in soft targets. For larger, higher-velocity calibers, such as tank guns, hardness is of secondary importance to density, and are normally sub-caliber projectiles made from tungsten carbide, tungsten hard alloy or depleted uranium fired in a light aluminum or magnesium alloy (or carbon fibre in some cases) sabot.

"Many modern tank guns are smoothbore, not rifled, because practical rifling twists can only stabilize projectiles, such as an armor-piercing fin-stabilized discarding sabot (APFSDS), with a length-to-diameter ratio of up to about 5:1, the spin imposed by rifling interferes with shaped-charge rounds, and also because the rifling adds friction and reduces the velocity it is possible to achieve. To get the maximum force on the smallest area,

anti-tank rounds have aspect ratios of 10:1 or more. Since these cannot be stabilized by rifling, they are built instead like large darts, with fins providing the stabilizing force, negating the need for rifling. These subcaliber rounds are held in place in the bore by sabots. The sabot is a light material that transfers the pressure of the charge to the penetrator, [and] then is discarded when the round leaves the barrel."

—Wikipedia, the free encyclopedia

Ballistic penetrating power (BPP) deals with the measurement of the bullet's power or ability to successfully penetrate a given target density at a given initial terminal velocity, and therefore represents the product of the depth of penetration (*Pd*) and the differential of the target density (*dTp*).

$$BPP = \frac{Pd}{dTp} \tag{1}$$

The differential, or the degree of difference, of target densities is the area of the spectrum of densities, from oxygen (0.0000509 lbs/cu./in.) to the pure metal rhenium (0.756 lbs/cu/.in.) in which a bullet will successfully penetrate.

If we were to increase the depth of penetration, in any given target and material, then we increase the BPP.

When we increase the target density, for a given bullet, we decrease the BPP.

CHAPTER 34

BALLISTIC WORK FUNCTION

"*In solid-state physics, the work function (sometimes spelled workfunction) is the minimum thermodynamic work (i.e. energy) needed to remove an electron from a solid to a point in the vacuum immediately outside the solid surface. Here "immediately" means that the final electron position is far from the surface on the atomic scale, but still too close to the solid to be influenced by ambient electric fields in the vacuum. The work function is not a characteristic of a bulk material, but rather a property of the surface of the material (depending on crystal face and contamination).*"

—Wikipedia, the free encyclopedia

For the science of small arms ballistics, we define *ballistic work function* (β) as the minimum amount of kinetic energy from a bullet we need to knock down a target, such as in competitive silhouette shooting events, without causing damage to that target, or penetrating it, with a given target and distance from the gun.

$$\beta = Wt \times Ds \quad (1)$$

where β is the product of target weight (*Wt*) and *Ds* is the distance from the gun.

When we increase the weight of the target, we increase the work we need to knock down the target, and when we increase the distance between the target and the gun, we also increase the amount of work we need to knock down the target.

BIBLIOGRAPHY

Ackley, P. O., (1962). *Handbook for Shooters and Reloaders*, volumes I and II, by Publishers Press, Salt Lake City.

Derek, B., (2004). Aeromechanical Systems Group, DAPS, RMCS, Cranfield University, Shrivenham, Swindon, Wiltshire, UK, *External Ballistics*.

Bliss, G. A., (1944). *Mathematics for Exterior Ballistics*, John Wiley & Sons, Inc., New York, New York.

Bradley, J. W., (1990). *An Alternative Form of the Modified Point-Mass Equation of Motion*, BRL Memorandum Report #3875, Ballistic Research Laboratories, Aberdeen Proving Ground, Maryland, November.

Carlucci, D., & Sidney, J., (2008). *Ballistics: Theory and Design of Guns and Ammunition*, CRC Press, Boca Raton, FL, ISBN 13:978-1-4200-6618-0 and 10:1-4200-6618-0.

Cloutier, M., (1972). *Methods for Computing Drag Coefficients from Ballistic Range Data Using the APL System*, , DRFV-TN-2027/72, Centre de Recherches Pour la Defense (Defense Research Establishment), Valcartier.

Corner, J., (1951). *Theory of the Interior Ballistics of Guns*, 1950, Wiley.

Cummings, C. S., (1950). *Everyday Ballistics*, 2nd Edn., Stackpole & Heck, Inc., New York, ASIN: B0007IXU6Q.

Dickinson, E. R., (1964). *Aerodynamic Effects of Blunting a Projectile Nose*, BRL Memorandum Report #1956, Ballistic Research Laboratories, Aberdeen Proving Ground, Maryland.

Duncan, L., & Engebos, B., (1967). *A Six-Degree-of-Freedom Digital Computer Program for Trajectory Simulation*, AD664116, Atmospheric Sciences Laboratory, White Sands Missile Range, New Mexico.

Farrar, C. L., & Leeming, D. W., (1983). *Military Ballistics: A Basic Manual*, 225 pages, 27.00 (ISBN 0.08.028342-X).

Hall, A. R., (1952). *Ballistics in the Seventeenth Century*, Cambridge.

Hager, W. W., DeJarnette, F. R., & Moore, F. G., (1977). *Optimal Projectile Shapes for Minimum Total Drag*, NSWC TR-79-18, Dalhgren, Virginia.

Harvard University, (1948). *Computation Laboratory Tables for the Design of Missiles*, Harvard University Press, Oxford.

Hatcher, J. S., (1962). *Hatcher's Notebook*, 3rd edition, Stackpole, ISBN 0811707954.

Hatcher, J. S., Wilhelm, G. P., & Malony, H. J., (1917). *Machine Guns*, Menasha, Wisc. George Banta Pub. Co., [Riling 1833].

Hatcher, J. S., (1927). *Pistols and Revolvers and Their Use*, Marshallton, Del., Small-Arms Technical Pub. Co., [Riling 2017].

Hatcher, J. S., (1935). *Textbook of Pistols and Revolvers*, Onslow County, N.C., Small-Arms Technical Pub. Co., [Riling 2170].

Hatcher, J. S., (1935). *Textbook of Firearms Investigation*, Plantersville, S.C., Small-Arms Technical Pub. Co., [Riling 2171].

Hatcher, J. S., (1947). *Hatcher's Notebook*, Harrisburg, Pa., Military Service Pub. Co., [Riling 2596].

Hatcher, J. S., (1947). *The Book of the Garand*, Washington, Infantry Journal Press, [Riling 2645].

Hatcher, J. S., Al Barr, H. P., White, C., & Neumann, L., *Handloading*, Washington, National Rifle Association of America, [Riling 2722].

Hayes, T. J., (1940). *Exterior Ballistics,* Wiley.

Hermann, E. E., (1930). *Exterior Ballistics*, [2d ed] 30 U.S. Naval Institute.

Hermann, E. E., (1930). *Range and Ballistic Tables*, [2d ed] U.S. Naval Institute.

Hitchcock, H. P., (1976). *Aerodynamic Data for Spinning Projectiles*, BRL Memorandum Report # 620, 1947, Ballistic Research Laboratories, Aberdeen Proving Ground, Maryland.

Hunt, F. R. W., et al. *Internal Ballistics,* 151 Philosophical Libraries, H.M.S.O.

Krasnov, N. F., Morris, & Deane, N., (1970). *Aerodynamics of Bodies of Revolution*, American Elsevier Publishing, New York, ISBN 0444000763.

Kooy, J. M. J., & Uytenbogaart, J. W. H., (1947). *Ballistics for the Future*, McGraw.

Laible, R. C., (1980). *Ballistic Materials and Penetration Mechanics* (Methods & Methods Ser: vol. 5), 64.00.

Lieske, R. F., & Danberg, J. E., (1992). *Modified Point Mass Trajectory Simulation for Base-Burn Projectiles*, , AD-A248 292, Ballistic Research Laboratories, Aberdeen Proving Ground, Maryland.

Lister, C. B., (1944). *Simplified Small Arms Ballistics*, NRA.

Litz, B., (2009). *Applied Ballistics for Long Range Shooting*, Applied Ballistics, LLC., ISBN 9780615276618.

Lowry, E. D., (1965). *Exterior Ballistics of Small Arms Projectiles,* Research Dept., Winchester-Western Division, Olin Mathieson Chemical Corporation, ASIN B0007FPINM.

MacNeish, H. F. *Elementary Mathematical Theory of Exterior Ballistics,* Brooklyn College Press, Bedford & Oxford, N.Y.

Mann, F. W., (1980). *The Bullet's Flight From Powder to Target – Ballistics of Small Arms*, Wolfe Pub Co; ISBN 0935632042.

Matunas, E., (1979). *American Ammunition & Ballistics*.

McCoy, R. L., (1999). *Modern Exterior Ballistics The Launch and Flight Dynamics of Symmetric Projectiles*, Schiffer Publishing, Ltd.; ISBN 0764307207.

McShane, E. J., et al. *Exterior Ballistics, 153* University of Denver Press.

McShane, K. R., (1953). *Exterior Ballistics*, University of Denver Press, ASIN: B0007E0YN2.

Moulton, F. R., (1962). *Methods in Exterior Ballistics*, Dover Publications, Inc., New York, ISBN 1114518514.

Moore, & Frank G., (1972). *Body Alone Aerodynamics of Guided and Unguided Projectiles at Subsonic, Transonic and Supersonic Mach Numbers*, Naval Weapons Laboratory, AD-754 098, Dahlgren, Virginia.

Rinker, R. A., (1995). *Understanding Ballistics*, Mulberry House Publishing, Corydon, IN, ISBN 0964559803.

United States Naval Academy, (1948). Department of Ordnance and Gunnery, *Exterior Ballistics*, U.S. Naval Institute.

United States Naval Academy, (1943). Department of Ordnance of Gunnery, *Exterior Ballistics*, 143 pages, U.S. Naval Institute.

INDEX

www.ingramcontent.com/pod-product-compliance
Ingram Content Group UK Ltd.
Pitfield, Milton Keynes, MK11 3LW, UK
UKHW022158190926
475702UK00003B/97
* 9 7 8 1 7 7 1 8 8 6 5 0 5 *